American National Standard Code for Electricity Metering

Published by
The Institute of Electrical and Electronics Engineers, Inc

Distributed in cooperation with
Wiley-Interscience, a division of John Wiley & Sons, Inc

ANSI C12.1-1982
(Revision of
ANSI C12-1965)

American National Standard Code for Electricity Metering

7th Edition

Approved July 6, 1982

American National Standards Institute

Secretariat

Institute of Electrical and Electronics Engineers, Inc
National Bureau of Standards

Published by

Institute of Electrical and Electronics Engineers, Inc
345 East 47th St, New York, NY 10017

ISBN 0-471-89356-0

Library of Congress Catalog Number 82-083450

© Copyright 1982 by

The Institute of Electrical and Electronics Engineers, Inc

February 18, 1983   *SH08623*

American National Standard

An American National Standard implies a consensus of those substantially concerned with its scope and provisions. An American National Standard is intended as a guide to aid the manufacturer, the consumer, and the general public. The existence of an American National Standard does not in any respect preclude anyone, whether he has approved the standard or not, from manufacturing, marketing, purchasing, or using products, processes, or procedures not conforming to the standard. American National Standards are subject to periodic review and users are cautioned to obtain the latest editions.

CAUTION NOTICE: This American National Standard may be revised or withdrawn at any time. The procedures of the American National Standards Institute require that action be taken to reaffirm, revise, or withdraw this standard no later than five years from the date of publication. Purchasers of American National Standards may receive current information on all standards by calling or writing the American National Standards Institute.

Foreword

(This Foreword is not a part of ANSI C12.1-1982, American National Standard Code for Electricity Metering.)

This standard has been enlarged to include performance specifications for a new class of self-contained watthour meters with increased load range. The specifications for other meters have been retained from the previous edition without major changes, but the presentation of some of the data has been rearranged to improve clarity. The section on standard watthour meters has also been revised to take account of the types of meters which have come into more widespread use during the last few years. Numerous other revisions are mainly editorial to correct errors and to bring the text into agreement with current standard terminology.

Since 1976, the C12 Committee has assumed responsibility for developing additional standards related to the Code for Electricity Metering, some of which were formerly issued by other organizations. By providing mechanical and other specifications, generally not directly related to performance, these standards complement the Code for Electricity Metering. The Code, which until this edition has been known as C12, has now been redesignated C12.1. The other standards issued by the C12 Committee are listed below.

C12.4-1978, American National Standard for Mechanical Demand Registers.

C12.5-1978, American National Standard for Thermal Demand Meters.

C12.6-1978, American National Standard for Marking and Arrangement of Terminal for Phase-Shifting Devices Used in Metering.

C12.7-1982, American National Standard Requirements for Watthour Meter Sockets.

C12.8-1981, American National Standard for Test Blocks and Cabinets for Installation of Self-Contained A-Base Watthour Meters.

C12.9-1982, American National Standard for Test Switches for Transformer-Rated Meters.

C12.10-1978, American National Standard for Watthour Meters.

C12.11-1978, American National Standard for Instrument Transformers for Metering Purposes, 15 kV and Less.

C12.14-1982, American National Standard for Magnetic Tape Pulse Recorders for Electricity Meters.

This standard was developed by the American National Standards Committee on Electricity Metering, C12, for full consensus approval as an American National Standard. Suggestions for improving this standard are welcome. They should be sent to the American National Standards Institute, 1430 Broadway, New York, NY 10018.

The Secretariat of the American National Standards Committee C12 on Electricity Metering is held by the Institute of Electrical and Electronics Engineers and the National Bureau of Standards. At the time this standard was processed and approved, the C12 Committee had the following members:

Subcommittee 4 — Acceptance of New Types of Watthour Meters

A. Fini, *Chairman*

J. Anderson	D. F. Becker
A. G. Ashenbeck, Jr	C. R. Collinsworth

Subcommittee 10 — Editorial

F. J. Levitsky, *Chairman*

C. F. Mueller
A. Loika
R. S. Turgel

In addition to the Committees listed above, ANSI C12 also has the following subcommittees:

Subcommittee		*Chairman*
Subcommittee 1	— Definitions	(*vacant*)
Subcommittee 2	— Measurement of Power and Energy	(*vacant*)
Subcommittee 5	— Watthour Test Method . . .	F. J. Levitsky
Subcommittee 6	— Installation Requirements .	(*vacant*)
Subcommittee 7	— Instrument Transformers and Auxiliary Devices	C. F. Mueller
Subcommittee 8	— In-Service Test for Watthour Meters	(*vacant*)
Subcommittee 9	— Demand Meters	C. R. Collinsworth
Subcommittee 11	— Safety Requirements	F. J. Levitsky
Subcommittee 12	— Solid-State Meters	R. S. Turgel
Subcommittee 13	— Time-of-Day Metering . . .	T. C. Drew
Subcommittee 14	— Pulse Recorders	T. C. Drew
Subcommittee 15	— Watthour Meter Sockets and Test Blocks	F. A. Marta

Contents

FIGURES

TABLES   PAGE

American National Standard Code for Electricity Metering

1. Scope

This Code establishes acceptable performance criteria for new types of ac watthour meters, demand meters, demand registers, pulse devices, instrument transformers, and auxiliary devices. It states acceptable in-service performance levels for meters and devices used in revenue metering. It also includes information on related subjects, such as recommended measurement standards, installation requirements, test methods, and test schedules. Some of these provisions are applicable to dc watthour meters as well, and acceptable in-service performance levels of such meters are given in Appendix F.

This Code for Electricity Metering is designed as a reference for those concerned with the art of electricity metering, such as utilities, manufacturers, and regulatory bodies.

2. Definitions

The definitions given apply specifically to the subject treated in this American National Standard. Most of them are grouped by general terms, such as *watthour meter*, and all are given numbers for identification.

For additional definitions see ANSI/IEEE Std 100-1977 [6][1], IEEE Standard Dictionary of Electrical and Electronics Terms.

2.1 accuracy. The extent to which a given measurement agrees with the defined value.

2.2 calibration. Comparison of the indication of the instrument under test, or registration of the meter under test, with an appropriate standard.

2.3 coupling-capacitor voltage transformer (CCVT). A voltage transformer comprised of a capacitor divider and an electromagnetic unit so designed and interconnected that the secondary voltage of the

---

[1]The numbers in brackets correspond to those of the references listed in Section 11.

electromagnetic unit is substantially proportional to, and in phase with, the primary voltage applied to the capacitor divider for all values of secondary burdens within the rating of the coupling-capacitor voltage transformer.

2.4 class designation. See 2.109.

2.5 contact-making clock. See 2.19.

2.6 creep. See 2.110.

2.7 crosstalk. Unwanted electric signals injected into a circuit by stray coupling.

2.8 current comparator. A device by which the ratio and phase angle between two currents can be measured precisely.

NOTE: A common form of current comparator relies on a balance of ampere-turns produced by currents in two or more windings on one or more magnetic cores.

2.9 current transformer. An instrument transformer designed for use in the measurement or control of current. See also 2.32.

NOTE: Its primary winding, which may be a single turn or bus bar, is connected in series with the load.

2.10 current transformer – continuous thermal current rating factor. The factor by which the rated primary current is multiplied to obtain the maximum allowable primary current based on the maximum permissible temperature rise on a continuous basis.

2.11 current transformer – phase angle. The angle between the current leaving the identified secondary terminal and the current entering the identified primary terminal. This angle is considered positive when the secondary current leads the primary current.

2.12 demand. The average value of power or a related quantity over a specified interval of time.

NOTE: Demand is expressed in kilowatts, kilovolt-amperes, kilovars, or other suitable units.

2.13 demand constant (pulse receiver). The value of the measured quantity for each received pulse, divided by the demand interval, expressed in kilowatts per pulse, kilovars per pulse, or other suitable units.

2.14 demand deviation. The difference between the indicated or recorded demand and the true demand, expressed as a percentage of the full-scale value of the demand meter or demand register. For pulse recorders see 2.61.

2.15 demand interval (block-interval demand meter). The specified interval of time on which a demand measurement is based. See also 2.23.

NOTE: Intervals such as 15, 30, or 60 minutes are commonly specified.

2.16 demand-interval deviation. The difference between the measured demand interval and the specified demand interval, expressed as a percentage of the specified demand interval.

2.17 demand – maximum. The highest demand measured over a selected period of time, such as 1 month.

2.18 demand meter. A metering device that indicates or records the demand, maximum demand, or both.

NOTE: Since demand involves both an electrical factor and a time factor, mechanisms responsive to each of these factors are required, as well as an indicating or recording mechanism. These mechanisms may be either separate from or structurally combined with one another.

2.19 demand meter – contact-making clock. A device designed to close momentarily an electric circuit to a demand meter at periodic intervals.

2.20 demand meter – indicating. A demand meter equipped with a readout that indicates demand, maximum demand, or both.

2.21 demand meter – integrating (block-interval). A meter that integrates power or a related quantity over a fixed time interval, and indicates or records the average.

2.22 demand meter – lagged. A meter that indicates demand by means of thermal or mechanical devices having an approximately exponential response.

2.23 demand meter – time characteristic (lagged-demand meter). The nominal time required for 90% of the final indication, with constant load suddenly applied.

NOTE: The time characteristic of lagged-demand meters describes the exponential response of the meter to the applied load. The response of the lagged-demand meter to the load is continuous and independent of the selected discrete time intervals.

2.24 demand meter – timing deviation. The difference between the elapsed time indicated by the timing element and the true elapsed time, expressed as a percentage of the true elapsed time.

2.25 demand register. A mechanism, for use with an integrating electricity meter, that indicates maximum demand and also registers electric energy (or other integrated quantity).

2.26 demand register – cumulative. A register that indicates the sum of the previous maximum demand readings prior to reset. When reset, the present reading is added to the previous accumulated readings. The maximum demand for the present reading period is the difference between the present and previous readings.

2.27 demand register – multiple-pointer form. An indicating demand register from which the demand is obtained by reading the position of the multiple pointers relative to their scale markings. The multiple pointers are resettable to zero.

2.28 demand register – single-pointer form. An indicating demand register from which the demand is obtained by reading the position of a pointer relative to the markings on a scale. The single pointer is resettable to zero.

2.29 electricity meter. A device that measures and registers the integral of an electrical quantity with respect to time.

2.30 energy. The integral of active power with respect to time.

2.31 inspection – meter installation. Examination of the meter, auxiliary devices, connections, and surrounding conditions, for the purpose of discovering mechanical defects or conditions that are likely to be detrimental to the accuracy of the installation. Such an examination may or may not include an approximate determination of the percentage registration of the meter.

2.32 instrument transformer. A transformer that reproduces in its secondary circuit, in a definite and known proportion, the voltage or current of its primary circuit, with the phase relation substantially preserved. See also 2.9 and 2.101.

2.33 instrument transformer – accuracy class. The limits of transformer correction factor, in terms of percent error, that have been established to cover specific performance ranges for line power factor conditions between 1.0 and 0.6 lag.

2.34 instrument transformer – accuracy rating for metering. The accuracy class, together with a standard burden for which the accuracy class applies.

2.35 instrument transformer – burden. The impedance of the circuit connected to the secondary winding.

NOTE: For voltage transformers it is convenient to express the burden in terms of the equivalent volt-amperes and power factor at a specified voltage and frequency.

2.36 instrument transformer – correction factor. The factor by which the reading of a wattmeter or the registration of a watthour meter must be multiplied to correct for the effects of the error in ratio and the phase angle of the instrument transformer. This factor is the product of the ratio and phase-angle correction factors for the existing conditions of operation.

2.37 instrument transformer – marked ratio. The ratio of the rated primary value to the rated secondary value as stated on the nameplate.

2.38 instrument transformer – phase angle. See 2.11 and 2.103.

2.39 instrument transformer – phase-angle correction factor. The factor by which the reading of a wattmeter or the registration of a watthour meter, operated from the secondary of a current transformer or a voltage transformer, or both, must be multiplied to correct for the effect of phase displacement of secondary current, or voltage, or both, with respect to primary values.

NOTE: This factor equals the ratio of true power factor to apparent power factor and is a function of both the phase angles of the instrument transformers and the power factor of the primary circuit being measured.

2.40 instrument transformer – ratio correction factor. The factor by which the marked ratio of a current transformer or a voltage transformer must be multiplied to obtain the true ratio.

NOTE: This factor is expressed as the ratio of true ratio to marked ratio. If both a current transformer and a voltage transformer are used in conjunction with a wattmeter or watthour meter, the resultant ratio correction factor is the product of the individual ratio correction factors.

2.41 instrument transformer – true ratio. The ratio of the magnitude of the primary quantity (voltage or current) to the magnitude of the corresponding secondary quantity.

2.42 laboratory – meter. A laboratory responsible for maintaining reference standards and assigning values to the working standards used for the testing of electricity meters and auxiliary devices.

2.43 laboratory – independent standards. A standards laboratory maintained by, and responsible to, a company or authority that is not under the same administrative control as the laboratories or companies submitting instruments for calibration.

2.44 maximum demand. See 2.17.

2.45 meter. See 2.18, 2.29, and 2.105.

2.46 meter laboratory. See 2.42.

21

2.47 meter shop. See 2.80.

2.48 percentage registration. See 2.120.

2.49 phase angle. See 2.11 and 2.103.

2.50 phase-angle correction factor. See 2.39.

2.51 phase-shifting transformer. See 2.98.

2.52 phasor. A complex number, associated with sinusoidally varying electrical quantities, such that the absolute value (modulus) of the complex number corresponds to either the peak amplitude or rms value of the quantity, and the phase (argument) to the phase angle at zero time. By extension, the term "phasor" can also be applied to impedance and related complex quantities that are not time dependent.

2.53 power – active. The time average of the instantaneous power over one period of the wave.

NOTE: For sinusoidal quantities in a two-wire circuit, it is the product of the voltage, the current, and the cosine of the phase angle between them. For nonsinusoidal quantities, it is the sum of all the harmonic components, each determined as above. In a polyphase circuit it is the sum of the active powers of the individual phases.

2.54 power – apparent. For sinusoidal quantities in either single-phase or polyphase circuits, apparent power is the square root of the sum of the squares of the active and reactive powers.

NOTE: This is, in general, not true for nonsinusoidal quantities.

2.55 power – reactive. For sinusoidal quantities in a two-wire circuit, reactive power is the product of the voltage, the current, and the sine of the phase angle between them. For nonsinusoidal quantities, it is the sum of all the harmonic components, each determined as above. In a polyphase circuit, it is the sum of the reactive powers of the individual phases.

2.56 power factor. The ratio of active power to the apparent power.

2.57 precision. The repeatability of measurement data, customarily expressed in terms of standard deviation.

2.58 pulse. A wave that departs from an initial level for a limited duration of time and ultimately returns to the original level.

NOTE: In demand metering, the term "pulse" is also applied to a sudden change of voltage or current, produced, for example, by the closing or opening of a contact.

2.59 pulse amplifier or relay. A device used to change the amplitude or waveform of a pulse for retransmission to another pulse device.

2.60 pulse capacity. The number of pulses per demand interval that a pulse receiver can accept and register without loss.

2.61 pulse-count deviation. The difference between the number of recorded pulses and the number of pulses supplied to the input terminals of a pulse recorder (true count), expressed as a percentage of the true count. Pulse-count deviation is applicable to each data channel of a pulse recorder.

2.62 pulse device (for electricity metering). The functional unit for initiating, transmitting, retransmitting, or receiving electric pulses, representing finite quantities, such as energy, normally transmitted from some form of electricity meter to a receiver unit.

2.63 pulse initiator. Any device, mechanical or electrical, used with a meter to initiate pulses, the number of which are proportional to the quantity being measured. It may include an external amplifier or auxiliary relay or both.

2.64 pulse-initiator coupling ratio. The number of revolutions of the pulse-initiating shaft for each output pulse.

2.65 pulse-initiator gear ratio. The ratio of meter rotor revolutions to revolutions of the pulse-initiating shaft.

2.66 pulse-initiator output constant. The value of the measured quantity for each outgoing pulse of a pulse initiator, expressed in kilowatt hours per pulse, kilovarhours per pulse, or other suitable units.

2.67 pulse-initiator output ratio. The number of revolutions of the meter rotor per output pulse of the pulse initiator.

2.68 pulse-initiator ratio. The ratio of revolutions of the first gear of the pulse initiator to revolutions of the pulse-initiating shaft.

2.69 pulse-initiator shaft reduction. The ratio of revolutions of the meter rotor to the revolutions of the first gear of the pulse initiator.

2.70 pulse rate – maximum. The number of pulses per second at which a pulse device is nominally rated.

2.71 pulse receiver. The unit that receives and registers the pulses. It may include a periodic resetting mechanism, so that a reading proportional to demand may be obtained.

23

2.72 pulse recorder. A device that receives and records pulses over a given demand interval.

NOTE: It may record pulses in a machine-translatable form on magnetic tape, paper tape, or other suitable media.

2.73 pulse-recorder channel. A means of conveying information. It consists of an individual input, output, and intervening circuitry required to record pulse data on the recording media.

2.74 pulse relay – totalizing. A device used to receive and totalize pulses from two or more sources for proportional transmission to another totalizing relay or to a receiver.

2.75 Q-hour meter. An electricity meter that measures the quantity obtained by effectively lagging the applied voltage to a watthour meter by 60 degrees. This quantity is one of the quantities used in calculating quadergy (varhours).

2.76 quadergy. The integral of reactive power with respect to time.

2.77 ratio correction factor. See 2.40.

2.78 registration. See 2.120.

2.79 response time (lagged-demand meter). See 2.23.

2.80 shop – meter. A place where meters are inspected, repaired, tested, and adjusted.

2.81 standards – basic reference. Those standards with which the values of the electrical units are maintained in the laboratory, and which serve as the starting point of the chain of sequential measurements carried out in the laboratory.

2.82 standards – dc-ac transfer. Instruments used to establish the equality of an rms current or voltage (or the average values of alternating power) with the corresponding steady-state dc quantity.

2.83 standards – laboratory reference. Standards that are used to assign and check the values of laboratory secondary standards.

2.84 standards – laboratory secondary. Standards that are used in the routine calibration tasks of the laboratory.

2.85 standards – national. Those standards of electrical measurements that are maintained by the National Bureau of Standards.

2.86 standards – transport. Standards of the same nominal value as the basic reference standards of a laboratory (and preferably of equal quality), which are regularly intercompared with the basic group but are reserved for periodic interlaboratory comparison tests to check the stability of the basic reference group.

2.87 test – acceptance. A test to demonstrate the degree of compliance of a device with the purchaser's requirement.

2.88 test amperes (TA). See 2.133.

2.89 test – approval. A test of one or more meters or other items under various controlled conditions to ascertain the performance characteristics of the type of which they are a sample.

2.90 test current. See 2.133.

2.91 test – in-service. A test made during the period that the meter is in service. It may be made on the customer's premises without removing the meter from its mounting, or by removing the meter for test, either on the premises or in a laboratory or meter shop.

2.92 test – referee. A test made by or in the presence of one or more representatives of a regulatory body or other impartial agency.

2.93 test – request. A test made at the request of a customer.

2.94 time characteristic (lagged-demand meter). See 2.23.

2.95 transducer. A device to receive energy from one system and supply energy, of either the same or of a different kind, to another system, in such a manner that the desired characteristics of the energy input appear at the output.

2.96 transformer. See 2.9, 2.32, and 2.101.

2.97 transformer-loss compensator. A passive electric network that adds to or subtracts from the meter registration to compensate for predetermined iron and copper losses of transformers and transmission lines.

2.98 transformer – phase-shifting. An assembly of one or more transformers intended to be connected across the phases of a polyphase circuit so as to provide voltages in the proper phase relations for energizing varmeters, varhour meters, or other measurement equipment. This type of transformer is sometimes referred to as a phasing transformer.

2.99 varhour meter. An electricity meter that measures and registers the integral, with respect to time, of the reactive power of the circuit in which it is connected. The unit in which this integral is measured is usually the kilovarhour.

2.100 varhour constant. The registration, expressed in varhours, corresponding to one revolution of the rotor.

25

2.101 voltage transformer. An instrument transformer designed for use in the measurement or control of voltage. See also 2.32.

NOTE: Its primary winding is connected across the supply circuit.

2.102 voltage transformer – continuous thermal burden rating. The volt-ampere burden that the voltage transformer will carry continuously at rated voltage and frequency without causing the specified temperature limitations to be exceeded.

2.103 voltage transformer – phase angle. The angle between the secondary voltage from the identified to the unidentified terminal and the corresponding primary voltage. This angle is considered positive when the secondary voltage leads the primary voltage.

2.104 voltage-withstand tests. Tests made to determine the ability of insulating materials and spacings to withstand specified overvoltages for a specified time without flashover or puncture.

2.105 watthour meter. An electricity meter that measures and registers the integral, with respect to time, of the active power of the circuit in which it is connected. This power integral is the energy delivered to the circuit during the interval over which the integration extends, and the unit in which it is measured is usually the kilowatthour.

2.106 watthour meter – adjustment. Adjustment of internal controls to bring the percentage registration of the meter to within specified limits.

2.107 watthour meter – basic current range. The current range of a multirange standard watthour meter designated by the manufacturer for the adjustment of the meter (normally the 5-ampere range).

2.108 watthour meter – basic voltage range. The voltage range of a multirange standard watthour meter designated by the manufacturer for the adjustment of the meter (normally the 120-volt range).

2.109 watthour meter – class designation. The maximum of the load range in amperes. See 2.117.

2.110 watthour meter – creep. A continuous motion of the rotor of a meter with normal operating voltage applied and the load terminals open-circuited.

2.111 watthour meter – form designation. An alphanumeric designation denoting the circuit arrangement for which the meter is applicable and its specific terminal arrangement. The same designation is applicable to equivalent meters of all manufacturers.

2.112 watthour meter – gear ratio. The number of revolutions of the rotor for one revolution of the first dial pointer, commonly denoted by the symbol R_g.

2.113 watthour meter – heavy load. See 2.133.

2.114 watthour meter – induction. A motor-type meter in which currents induced in the rotor interact with a magnetic field to produce the driving torque.

2.115 watthour meter – light load. The current at which the meter is adjusted to bring its response near the lower end of the load range to the desired value. It is usually 10% of the test current for a revenue meter and 25% for a standard meter.

2.116 watthour meter – load current. See 2.133.

2.117 watthour meter – load range. The range in amperes over which the meter is designed to operate continuously with specified accuracy.

2.118 watthour meter – motor type. A motor in which the speed of the rotor is proportional to the power, with a readout device that counts the revolutions of the rotor.

2.119 watthour meter – percentage error. The difference between its percentage registration and 100%. A meter whose percentage registration is 95% is said to be 5% slow, or its error is –5%. A meter whose percentage registration is 105% is 5% fast, or its error is +5%.

2.120 watthour meter – percentage registration. The percentage registration of a meter is the ratio of the actual registration of the meter to the true value of the quantity measured in a given time, expressed as a percentage.

2.121 watthour meter – portable standard. A portable meter, principally used as a standard for testing other meters. It is usually provided with several current and voltage ranges and with a readout indicating revolutions and fractions of a revolution of the rotor.

NOTE: Electronic portable standards not using a rotor may have a readout indicating equivalent revolutions and fractions of revolutions, or other units such as percentage registration.

2.122 watthour meter – rated current. The nameplate current for each range of a standard watthour meter.

NOTE: The main adjustment of the meter is ordinarily made with rated current on the basic current range.

2.123 watthour meter – rated voltage. The nameplate voltage for a meter or for each range of a standard watthour meter.

NOTE: The main adjustment of the standard meter is ordinarily made with rated voltage on the basic voltage range.

2.124 watthour meter – reference performance. Performance at specified reference conditions for each test, used as a basis for comparison with performance under other conditions of the test.

2.125 watthour meter – reference standard. A meter used to maintain the unit of electric energy. It is usually designed and operated to obtain the highest accuracy and stability in a controlled laboratory environment.

2.126 watthour meter – register. That part of the meter that registers the revolutions of the rotor, or the number of pulses received from or transmitted to a meter, in terms of units of electric energy or other quantity measured.

2.127 watthour meter – register constant. The multiplier used to convert the register reading to kilowatthours (or other suitable units).

NOTE: This constant, commonly denoted by the symbol, K_r, takes into consideration the watthour constant, gear ratio, and instrument transformer ratios.

2.128 watthour meter – register ratio. The number of revolutions of the first gear of the register, for one revolution of the first dial pointer.

NOTE: This is commonly denoted by the symbol, R_r.

2.129 watthour meter – registration. The registration of a meter is the apparent amount of electric energy (or other quantity being measured) that has passed through the meter, as shown by the register reading. It is equal to the product of the register reading and the register constant. The registration during a given period is equal to the product of the register constant and the difference between the register readings at the beginning and the end of the period.

2.130 watthour meter – rotor. That part of the meter that is directly driven by electromagnetic action.

2.131 watthour meter – standard. See 2.121 and 2.125.

2.132 watthour meter – stator. An assembly of an induction watthour meter, which consists of a voltage circuit, one or more current circuits, and a combined magnetic circuit so arranged that their joint effect, when energized, is to exert a driving torque on the rotor by the reaction with currents induced in an individual or common conducting disk.

2.133 watthour meter – test current (TA). The current specified by the manufacturer for the main adjustment of the meter (heavy- or full-load adjustment). See 8.1.3.2.

NOTES: (1) It has been identified as "TA" on revenue meters manufactured since 1960.

(2) The main adjustment of a meter used with a current transformer may be made either at the test current or at the rated secondary current of the transformer.

2.134 watthour meter – two-rate. A meter having two sets of register dials, with a changeover arrangement such that integration of the quantity will be registered on one set of dials during a specified time each day, and on the other set of dials for the remaining time.

2.135 watthour meter – watthour constant. The registration, expressed in watthours, corresponding to one revolution of the rotor.

NOTE: It is commonly denoted by the symbol K_h. When a meter is used with instrument transformers, the watthour constant is expressed in terms of primary watthours. For a secondary test of such a meter, the constant is the primary watthour constant divided by the product of the nominal ratios of transformation.

2.136 waveform distortion – percent. The ratio of the root-mean-square value of the harmonic content (excluding the fundamental) to the root-mean-square value of the nonsinusoidal quantity, expressed as a percentage.

3. Measurement of Power, Energy, and Related Quantities

3.1 Measurement of Power

3.1.1 Introduction. The growth in the use of electric power and energy has made necessary the adoption of polyphase alternating-current transmission and distribution systems. Such a system is a circuit or network to which are applied two or more voltages of the same frequency but displaced in phase by a fixed amount relative to one another. The individual circuits making up the polyphase network are called phases. The correct measurement of power, energy, and related quantities in polyphase circuits requires the proper selection and application of meters and meter elements.

It is not the intention to present in this standard a complete text of all the methods of measurement of power and energy. The material contained herein is intended to cover the basic methods of measurement of power, energy, and related quantities as accomplished by acceptable commercial practice.

3.1.2 Blondel's Theorem. In any system of N wires the true power may be measured by connecting a wattmeter in each line except one ($N - 1$ wattmeters), the current coil being in series with the line and the voltage coil connected between that line and the line containing no current coil. The total power for any load condition is the algebraic sum of the readings of all wattmeters so connected, provided that a grounded neutral connection to load or source is held equivalent to the addition of another wire.

All methods of measuring power or energy discussed in the remainder of Section 3 constitute applications of Blondel's theorem, with the foregoing limitations.

3.1.3 Direct-Current Circuits

3.1.3.1 Two-Wire DC Circuits. The total power in a two-wire direct-current circuit may be measured by one wattmeter. On grounded circuits, the current coil of the wattmeter must be connected in the ungrounded side of the circuit. It is generally preferable to connect the voltage coil leads across the terminals of the receiving circuit and, if greater accuracy is desired, the wattmeter reading may be corrected for its voltage-coil losses. If it is desired to minimize any possible effect of stray fields, the mean of reversed readings should be used.

3.1.3.2 Three-Wire DC Circuits. The total power in a three-wire direct-current circuit may be measured by two wattmeters, the current coils being connected one in each of the outside wires and the voltage leads of each wattmeter connected between its current coil wire, preferably on the receiving circuit side, and the third, common or neutral wire.

3.1.3.3 Ammeter and Voltmeter Method. In direct-current circuits an ammeter and a voltmeter may be used in place of a wattmeter. The product of their readings is the total power in the receiving circuit in which the instruments are connected, under steady-state conditions.

3.1.4 Single-Phase Alternating-Current Circuits

3.1.4.1 Single-Phase Two-Wire Circuits. The total power in a single-phase two-wire circuit may be measured by one wattmeter connected as in 3.1.3.1 except that the mean of reversed readings is not necessary.

3.1.4.2 Single-Phase Three-Wire Circuits. The total power in a single-phase three-wire circuit may be measured by two wattmeters connected as in 3.1.3.2.

3.1.5 Two-Phase Circuits

3.1.5.1 Two-Phase Three-Wire Circuits. The power in a two-phase three-wire circuit may be measured by two wattmeters, the current coils being connected one in each of the phase conductors and the voltage coil of each wattmeter connected between its current coil conductor and the common return. This method is correct for all conditions of load.

3.1.5.2 Two-Phase Four-Wire Circuits. The power in a two-phase four-wire circuit may be measured by two wattmeters, one of which is connected in each of the two phases as in 3.1.4.1. This method is correct for all conditions of load, provided that the midpoints of the two phases are not interconnected or grounded.

3.1.5.3 Two-Phase Five-Wire Circuits. The power in a two-phase five-wire circuit may be measured by means of four wattmeters, each having its current coils connected one in each of the phase conductors and the voltage coils connected between the corresponding phase conductor and the common conductor, or the neutral. This method is correct for all conditions of load.

3.1.5.4 Balanced Two-Phase Circuits. The power in a balanced two-phase three- or four-wire circuit may be measured by connecting a wattmeter in one phase, as in 3.1.4.1, and multiplying its readings by two.

3.1.6 Three-Phase Circuits

3.1.6.1 Three-Wattmeter Method. If the three loads are accessible as single-phase two-wire loads, the total power may be measured as the sum of the readings of the three wattmeters, each connected to one of the three loads as described in 3.1.4.1. This method is correct for all conditions of loading. This method is also correct for three-phase four-wire circuits, except that the voltage coil of each watt-meter is connected between the line conductor in which its current coil is connected and the common conductor, or the neutral.

3.1.6.2 Two-Wattmeter Method. The total power in a three-phase three-wire circuit may be measured by means of two wattmeters, having the current coils connected one in each of two line conduc-tors and the voltage coils connected between the line conductor in which its current coil is connected and the third line conductor. The algebraic sum of the readings of the two wattmeters indicates the total power supplied to any type of loading on the three con-ductors. This method is correct for any balanced or unbalanced load and for any power factor, but does not apply to three-phase four-wire circuits.

3.1.6.3 Balanced Three-Phase Circuits. The power in a balanced three-phase three- or four-wire wye circuit may be measured by one wattmeter by connecting its current coil in one phase conductor and its voltage coil between that conductor and neutral, real or artificial, and multiplying its readings by three.

3.2 Measurement of Energy

3.2.1 Basic Considerations. In general, electric energy is measured in the same way as electric power, by substituting an integrating watthour meter for a wattmeter. However, under certain circum-stances, for economic reasons, slight departures from Blondel's theorem are permissible, depending upon the degree of unbalance between two or more of the voltages in a multivoltage circuit. Care must be exercised concerning the degree to which those conditions are met in any specific case, in order to ensure that the accuracy of the metering lies within acceptable metering practice. An acceptable limit of metering accuracy or voltage unbalance may be defined as one in which the metering device reliably registers the electrical quantity that passes through it, in compliance with the require-ments of this standard. (See Section 5, Acceptable Performance of New Types of Electricity Meters and Instrument Transformers.)

In referring to watthour meters, it is convenient to refer to each complete electromagnetic structure containing current and voltage windings as a stator, each such stator being comparable to a single-phase wattmeter. Each such stator may consist of a voltage coil and a single current coil, or a voltage coil and two current coils. The single-stator two-wire meter has one voltage coil and one current coil. The single-stator three-wire meter has one voltage coil and two current coils, each of the latter having one-half the number of turns of the current coil in a two-wire meter. Meters may have one or more stators, each of which may be of the two-wire or three-wire type.

3.2.2 Direct-Current or Single-Phase Circuits

3.2.2.1 Two-Wire DC and Single-Phase Circuits. The energy in a two-wire direct-current or single-phase circuit may be measured by means of one watthour meter. If the circuit is grounded, the current coil should be connected in the ungrounded side of the circuit. Although in the measurement of power it is generally preferable to connect the voltage-coil leads across the terminals of the receiving circuit, as indicated in 3.1.3.1, for measurement of energy it is standard practice to connect the voltage-coil leads on the line side of the meter current coils.

3.2.2.2 Three-Wire DC and Single-Phase Circuits. The energy in a three-wire direct-current or single-phase circuit may be measured by two watthour meters, dc or ac, respectively. The meters are connected as were the wattmeters in 3.1.3.2 and 3.1.4.2, respectively. The total energy is the sum of the registrations of the two meters. The two meters may be arranged in a single housing and their combined registration recorded on a single register as in a two-stator meter.

3.2.2.3 Single-Phase Three-Wire Circuits with Balanced Voltages. A departure from Blondel's theorem may be used in a three-wire direct-current or single-phase circuit, provided that the voltages are balanced within acceptable limits. Under these conditions a single-stator three-wire meter may be used, having one voltage coil connected between the two ungrounded wires, and a two-section current winding consisting of two coils wound in opposite directions on a common core. Thus, when each of the current coils is connected in series with each of the line wires of the three-wire circuit, the magnetic effects of the currents flowing in the two coils are additive. The total number of turns of the two coils is the same as would be used on a single-winding two-wire stator of the same current rating. The accuracy of this method is independent of current or power factor balance, but it is dependent on voltage balance.

3.2.2.4 Large-Capacity or High-Voltage DC Meters. In direct-current circuits carrying heavy currents, shunt-type meters may be used, and in circuits operating above 240 V, resistors are used in the voltage-coil circuit of the meter to reduce the values of current and voltage applied to the meter in known and definite ratios that bring the current and voltage within the range of a meter having normal current and voltage ratings.

3.2.2.5 Large-Capacity or High-Voltage AC Meters. In alternating-current circuits carrying heavy currents or operating at high voltages, or both, current and voltage transformers are used to reduce the value of current and voltage, respectively, applied to the meter in known and definite ratios that bring the current and voltage within the range of a meter having normal current and voltage ratings. Current transformers should be used in all high-voltage metered circuits for purposes of insulation and safety, regardless of whether the value of the current necessitates their use.

3.2.3 Open Wye Circuits. The measurement of energy in a three-wire 120/208 V distribution circuit (two lines and a neutral) obtained from a four-wire wye three-phase system requires a two-stator meter to satisfy Blondel's theorem.

A departure from this theorem may be used to meter this circuit, provided that the line-to-neutral voltages are balanced in magnitude and phase angle within acceptable limits. There are two basic designs of a single-stator meter, called a network meter, that may be used for this purpose. Both designs utilize one voltage coil and two current coils. Depending on whether the voltage coil is energized from the line-to-neutral or the line-to-line voltage, one or two phase-shifting networks are employed to shift the phase of the current in one or both of the current coils in the proper amount and in the right direction to enable them magnetically to react correctly with the line-to-neutral or line-to-line voltage, respectively. In the design utilizing a voltage coil energized by line-to-line voltage, the number of voltage-coil turns is reduced from that of a 240 V coil to compensate for the reduction in voltage from 240 V to 208 V. Either meter will register correctly regardless of how the loads are connected or what the individual power factors of these loads may be, provided that the voltages are balanced and symmetrical, and in the correct phase sequence. Since a particular phase sequence is essential to the correct registration of this meter, a visual phase-sequence indicator is a built-in feature.

3.2.4 Two-Phase Circuits

3.2.4.1 Two-Phase Three-Wire Circuits. The energy in a two-phase three-wire circuit may be metered by means of two watthour meters connected in the same way as the two wattmeters described in 3.1.5.1.

3.2.4.2 Two-Phase Four- or Five-Wire Circuits. The energy in a two-phase four- or five-wire circuit may be metered by substituting a watthour meter for each wattmeter specified in 3.1.5.2 and 3.1.5.3, and any two or more meters can be combined in a single housing and their combined registration recorded on a single register, as described in 3.2.2.2.

3.2.4.3 Two-Phase Five-Wire Circuits with Balanced Voltages. A departure from Blondel's theorem may be used in the two-phase five-wire circuit if the voltages are balanced within acceptable limits. Two single-stator three-wire single-phase watthour meters, as described in 3.2.2.3, may be used. In this case the current coils of one meter are inserted, one in each of the phase wires of one of the two phases, and the voltage coil of the meter is connected between the same two wires. The second meter is connected similarly in the other pair of phase wires. The accuracy of this method is independent of current and power factor balance, but is dependent on voltage balance. The two three-wire meters may be combined in a single housing and their combined registration recorded on a single register.

3.2.5 Three-Phase Circuits, All Types

3.2.5.1 Methods Comparable to Power Measurements. Using watthour meters, the energy in any form of a three-phase circuit may be metered in accordance with the methods specified for power measurements by means of wattmeters described in 3.1.6.1, 3.1.6.2, or 3.1.6.3.

3.2.6 Three-Phase Three-Wire Circuits

3.2.6.1 Two-Stator Method. The energy in a three-phase three-wire circuit may be metered by means of a two-stator meter having its current coils connected in any two of the line conductors and the voltage coils connected between the line conductors in which the current coils are connected and the third line conductor. This method is correct for all conditions of load or voltage balance or unbalance and for any power factor.

3.2.6.2 Three-Stator Method. The energy in a three-phase three-wire circuit may be metered by means of three single-stator meters, or with one three-stator meter by establishing an artificial wye neutral for obtaining the phase voltages. Delta-wye-connected voltage transformers may be used for establishing these voltages.

3.2.6.3 Balanced Voltage and Load. In addition to the methods mentioned in 3.2.5.1, 3.2.6.1, and 3.2.6.2, and under the conditions of balanced voltage and balanced load of any power factor (as long as the power factor is the same in all phases), energy in a three-phase three-wire system may be measured by a single-stator three-wire watthour meter, having its current coils connected in any two of the line conductors and its voltage coil across the same two conductors, and multiplying its reading by two. This method is useful in measuring energy used by motors or other loads having balanced characteristics.

3.2.7 Three-Phase Four-Wire Wye Circuits

3.2.7.1 Three-Stator Four-Wire Wye Meter. The energy in a three-phase four-wire wye-connected circuit may be metered by a three-stator wye meter. The common point of the voltage circuits should be connected to the neutral conductor. This method is accurate for all conditions (balanced or unbalanced) of load, power factor, or voltage.

3.2.7.2 Two-Stator Four-Wire Wye Meter. The energy in a three-phase four-wire wye-connected circuit may be metered by a two-stator four-wire wye meter. According to Blondel's theorem, such a circuit would require a three-stator meter as described in 3.2.7.1. However, if the voltages between each line and neutral are balanced within acceptable limits, the accuracies generally are considered to be satisfactory. Such a meter has one voltage coil and a two-section current winding on each stator. This winding consists of two coils wound in opposite directions on a common core. Thus when each of the coils is connected in its respective circuit, the magnetic effects of the currents in the two sections of the winding are additive. These windings are connected as follows: One current coil of the first stator is inserted in one line conductor and its flux reacts with the flux of the voltage coil connected between that conductor and the neutral. Similarly, one current coil of the second stator is inserted in another line conductor and its flux reacts with the flux of the voltage coil connected between that conductor and the neutral. The remaining current coils, one on each stator, are connected in series and inserted in the remaining phase conductor.

3.2.8 Three-Phase Four-Wire Delta Circuits

3.2.8.1 Possible Methods of Metering. The energy in a three-phase four-wire open or closed delta-connected circuit, with the neutral formed by a tap to the midpoint of one of the phase windings, may be metered either by a three-stator meter or by a two-stator four-wire delta meter in several arrangements, provided that, in the latter case, the tapped phase voltages are balanced within acceptable limits.

3.2.8.2 Three-Stator Four-Wire Delta Meter. In a three-stator meter for three-phase four-wire delta circuits, one of the stators may have one-half the current and twice the voltage rating of the other two stators, but all three stators may be of the same rating provided that current transformers and voltage transformers of suitable different ratios are connected into the circuit. All three stators may also have the same ratings if the meter is properly calibrated to the voltages that are to be used in service, or if modern voltage coils are employed having an acceptable operating range that includes all voltages to be used.

The two current coils of equal rating are connected one each in the two phase conductors that have the mid-tap between them and their associated voltage coils connected between the corresponding phase conductors and the mid-tap or neutral conductor. The one-half-rated current coil is connected in the remaining phase conductor and its double-rated voltage coil is connected between that phase conductor and the mid-tap (neutral). This method is accurate for all conditions of loading and power factor with or without voltage balance.

3.2.8.3 Two-Stator Four-Wire Delta Meter. If the mid-tap is a true mid-tap (namely, the voltages from it to each associated line conductor are equal within acceptable limits), then a two-stator meter may be used, one stator of which has a three-wire current coil or the instrument transformer equivalent thereof. The two current coils of the three-wire stator are connected one each in the two line conductors having the mid-tap between them and their voltage coil connected between these same two line conductors. The current coil of the two-wire stator is connected in the third line conductor and its associated voltage coil between that conductor and the mid-tap (neutral).

3.2.9 Three-Phase Seven-Wire Double Wye-Connected Circuits

3.2.9.1 Three Single-Phase Three-Wire Stators. The energy in a three-phase seven-wire double wye-connected circuit may be measured by means of three single-phase three-wire watthour meters, or their polyphase equivalent. This method is correct for all values of balanced or unbalanced current and power factor, provided that the voltages are symmetrical and balanced within acceptable limits.

3.2.10 Basic Meter Design Considerations. The designers and manufacturers of induction watthour meters have always placed great emphasis on attaining as high a degree of inherent accuracy as was economically feasible. Induction watthour meters using modern materials and techniques are designed to function satisfactorily over a very wide load range. For these meters the terms nominal or rated load, or some multiple or fraction thereof, have no specific meaning. As a result, present-day practice classifies such a meter as Class 100 or Class 200. This means that a Class 100 meter is designed to operate continuously with acceptable accuracy up to a maximum current of 100 amperes, and a Class 200 meter to 200 amperes.

Since, however, this maximum current is not a suitable value to use when calibrating or testing a meter, the manufacturer designates the recommended value of amperes, called the test amperes, to be used when calibrating the meter. For example, a Class 100 meter might have a test-ampere designation of 15 amperes, abbreviated as TA 15, and a Class 200 meter might have a TA 30 nameplate rating.

Because modern meters are frequently required to perform with acceptable accuracies at values of current, voltage, frequency, etc, that may differ appreciably from those used to calibrate the meter, compensating devices have been developed to maintain, within acceptable limits, the calibration accuracy at the calibrating points, and over wide variations therefrom. Moreover, such devices are used to compensate for environmental conditions, such as changes in ambient temperature, and for other conditions that are not always ideal. No compensating device is perfect, but all in current use perform well within acceptable limits.

3.2.11 Factors Affecting Meter Accuracy

3.2.11.1 Light Loads. Due to certain nonlinear properties of even the best magnetic materials, the accuracy at very light loads (10% of TA value or less) would not be generally satisfactory if a device were not provided to correct for this nonlinearity. This is the primary function of the so-called light-load adjustment. When properly set, the accuracy at even very light loads is well within acceptable limits. In addition, this device will also compensate for any constant inherent friction or for any excess friction that might result from requiring the meter to drive auxiliary equipment such as pulse devices.

3.2.11.2 Variations in Voltage. The configuration of the electromagnetic voltage circuit in the modern induction meter is such as to minimize the error due to variations in voltage. Even with variations as large as 50% less than rated voltage, the error is within acceptable limits, and for all variations of the order of $\pm$ 10% to $\pm$ 15% the variation in accuracy is usually negligible.

3.2.11.3 Variations in Power Factor. An induction meter registers correctly at power factors less than unity only when a specific phase relation exists between certain of the torque-producing voltages, currents, and fluxes. This specific phase relationship is attained by means of a plate or coil, which initially can be adjusted by a calibration procedure. When correctly set, the accuracy of the meter, even at low power factors, is well within acceptable requirements. If this calibration is made with a lagging power factor and the meter is then operated at a leading power factor, a slight difference in its accuracy, usually negligible, may result.

3.2.11.4 Variations in Frequency. Frequency variations in a modern power system under normal operating conditions are insignificant. Any inaccuracies that might result from such variations as do exist are entirely negligible.

3.2.11.5 Variations in Temperature. The modern practice of placing meters outdoors has subjected such meters to a wide range of ambient temperatures. Fortunately, before this practice had become general, means of compensating the induction meter for the detrimental effects of changes in ambient temperature had been developed and incorporated in all modern meters. It may be assumed with confidence that all modern meters will function satisfactorily under all reasonable variations in ambient temperature.

3.2.11.6 External Magnetic Fields. In all well-designed induction watthour meters, the arrangement, number, and configuration of the various electromagnet and permanent-magnet circuits, as well as the number and arrangement of the several coils, are such as to keep the detrimental effect of external magnetic fields to a minimum. However, care should be exercised not to place the meter in a strong varying magnetic field of the same frequency as the rated frequency of the meter.

3.2.11.7 Load Range. One of the most unique features of the modern induction watthour meter is its ability to accurately measure loads many times its test-ampere rating. A load-compensating means is a built-in feature of all modern meters. With this device the test-ampere-calibration accuracy is maintained within very close limits up to and including the class ampere value. For both accuracy and thermal reasons, care should be exercised not to exceed the class value for any appreciable length of time.

3.2.11.8 Surges. Meters installed in rural areas are more exposed to the elements than are those in the more congested urban areas. As a result, atmospheric electrical disturbances have a greater opportunity to affect adversely the proper functioning of such meters.

These disturbances are capable of producing, under certain circumstances, very large currents of extremely short duration, called surge currents. These currents may go to ground through or in the vicinity of the meter. When this happens, the excessively large magnetic field created may affect the strength of the permanent magnets in the meter, thereby resulting in registration errors.

Present-day permanent magnets are designed to have a very strong ability to resist demagnetization. Thus, with modern meters, over-registration caused by surge currents is a rare occurrence.

In addition to surge proofing of the permanent magnets, a modern meter has built-in surge proofing for its insulation. For both voltage and current coils, built-in protective gaps to ground are used. In addition, the potential coil has increased surge resistance across the coil.

3.2.11.9 Adverse Environmental Conditions. Another result of installing a meter outdoors without any protective cover, other than its own enclosure, is that the meter may be exposed to dust and corrosive atmospheres such as salt spray.

Modern meters have been designed to minimize the detrimental effect of such exposures. Better sealing techniques, elimination of dissimilar metals, the use of protective coatings, stainless steels, and anodizing of aluminum are some of the precautions taken to counter the effect of adverse environmental conditions.

3.3 Measurement of Power Factor

3.3.1 Single-Phase Two-Wire Circuits. When the power is measured by one wattmeter in accordance with 3.1.4.1, power factor may be determined by connecting an additional wattmeter with its current coil in series with that of the first wattmeter, and with its voltage coil connected to an equal voltage displaced 90 degrees from that applied to the wattmeter connected for the power measurement. The additional wattmeter then measures reactive power in vars. Instruments in which the required 90-degree shift in phase is provided internally (or by means of accessories) are known as varmeters. From the measurements so obtained the power factor may be determined from the following formula for sinusoidal quantities:

$$\text{Power factor} = \frac{P}{\sqrt{P^2 + Q^2}}$$

where
   Q = reactive power
   P = active power

3.3.2 Single-Phase and Polyphase Circuits. When the power is measured by one or more wattmeters in accordance with 3.1.4, 3.1.5, or 3.1.6.1, the power factor of each measured separate phase is equal to the power in watts indicated by each wattmeter divided by the product of the voltage across the voltage circuit of the wattmeter and the current in the current coil of the wattmeter.

3.3.3 Balanced Three-Phase Three-Wire Circuits. In the case of a balanced three-phase three-wire circuit, where the power is measured by two wattmeters (see 3.1.6.2), the readings of the two wattmeters will be unequal when the power factor is less than unity, and if the power factor is less than 0.5 lag, the reading of one wattmeter will be negative.

3.3.4 System Power Factor. When both voltages and currents are balanced in any three-phase circuit, the system power factor may be determined from the following formula:

$$\text{Power factor} = \frac{P}{\sqrt{3}\,EI}$$

where

P = active power
E = line-to-line voltage
I = line current

3.3.5 Interval Power Factor. When a polyphase system is unbalanced in any manner, the system power factor ceases to have a specific physical meaning, such as the cosine of some particular phase angle. However, a numerical ratio can be obtained, called the interval power factor, defined as follows:

$$\text{Interval power factor} = \frac{Pt}{\sqrt{(Pt)^2 + (Qt)^2}}$$

where

Qt = product of total reactive power and time (as measured by a varhour meter)
Pt = product of total active power and time (as measured by a watthour meter)

The quantity thus defined is not in general equal to the average value of the power factor during the interval, but may be referred to as the interval power factor.

3.4 Measurement of Reactive Energy (Quadergy). Single-phase quadergy (varhours) may be measured with a varhour meter, usually a watthour meter with the current through the voltage coil displaced 90 degrees in phase from normal. A polyphase varhour meter, or an unmodified polyphase watthour meter used with an appropriate phase-shifting transformer, may be used to measure polyphase quadergy. If both leading and lagging quadergy are to be measured, two such meters with opposite displacements, and equipped with detents to prevent backward rotation, are usually required. To avoid this, a polyphase Q-hour meter[2], in which the voltages are lagged by 60 degrees either internally or by external connections to the polyphase circuit, can be used, along with a polyphase watthour meter. The Q-hour meter rotates in its normal direction for load phase angles from 30 degrees lead to 150 degrees lag. Quadergy in varhours, for a balanced-voltage system, can be computed from the meter readings with the formula:

$$\text{varhours} = \frac{(Q\text{-hours}) - (\text{watthours} \cdot \cos 60°)}{\sin 60°}$$

---

[2]The term "Q-hour" is in common usage, but should not be confused with the varhour, or Qt in 3.3.5.

4. Standards and Standardizing Equipment

4.1 General. The purpose and scope of this section is to specify the standards of electrical measurement and of time interval to which the metering of electric energy shall be referred, and to outline an appropriate chain of intermediate steps between the national standards of measurement and the watthour meters used in the meter shop.

4.2 Final Authority. The duties of the National Bureau of Standards, US Department of Commerce, include the following function assigned by Public Law 619 of the 81st Congress (64 Stat 371, 5 USC 271-286): "The custody, maintenance, and development of the national standards of measurements, and the provision of means and methods for making measurements consistent with these standards." In addition, Public Law 617 (Section 12) of the 81st Congress (Title 15, USCA Sec 223, 224) reads as follows: "It shall be the duty of the National Bureau of Standards to establish the values of the primary electric units in absolute measure, and the legal values for these units shall be those represented by, or derived from, national reference standards maintained by the National Bureau of Standards."

4.2.1 Electrical Units. The present units, as stated in Public Law 617 of the 81st Congress (Title 15, USCA Sec 223, 224) are defined in terms of the cgs (centimeter-gram-second) electromagnetic units,[3] as follows:

4.2.1.1 "The unit of electrical resistance shall be the ohm, which is equal to one thousand million units of resistance in the centimeter-gram-second system of electromagnetic units."

4.2.1.2 "The unit of electric current shall be the ampere, which

---

[3]Before January 1, 1948, the legal electrical units, defined in Public Law 105 of the 53rd Congress (28 Stat, Ch 131, p 102), were based on the resistance of a specified column of mercury measured under specified conditions, and on an unvarying current that would deposit silver at a specified rate from a silver nitrate solution under specified conditions. These defined what were at that time called the "International Ohm" and "International Ampere"; and the remaining "International Electrical Units" — of voltage, power, energy, etc — were defined in terms of known interrelations of these and the mechanical units of time and distance. The "International" units were superseded on January 1, 1948, by the "Absolute" units defined in 4.2.1 of this standard, which differ from the previous legal units by small amounts. The so-called "International Units" based on the mercury ohm and silver ampere must not be confused with the present "Système International d'Unites" (abbreviated SI units) approved by the International Electrotechnical Commission and officially adopted for use in the United States. This latter system is based on the meter, kilogram, second, and ampere, and the electrical units it provides are identical in magnitude with the "Absolute" units defined in 4.2.1.

is one-tenth of the unit of current in the centimeter-gram-second system of electromagnetic units."

4.2.1.3 "The unit of electromotive force [EMF] and of electric potential shall be the volt, which is the electromotive force that, steadily applied to a conductor whose resistance is one ohm, will produce a current of one ampere."

4.2.1.4 "The unit of electric quantity shall be the coulomb, which is the quantity of electricity transferred by a current of one ampere in one second."

4.2.1.5 "The unit of electrical capacitance shall be the farad, which is the capacitance of a capacitor which is charged to a potential of one volt by one coulomb of electricity."

4.2.1.6 "The unit of electrical inductance shall be the henry, which is the inductance in a circuit such that an electromotive force of one volt is induced in the circuit by variation of an inducing current at the rate of one ampere per second."

4.2.1.7 "The unit of power shall be the watt, which is equal to ten million units of power in the centimeter-gram-second system, and which is the power[4] required to cause an unvarying current of one ampere to flow between points differing in potential by one volt."

4.2.1.8 "The units of energy shall be (a) the joule, which is equivalent to the energy supplied by a power of one watt operating for one second, and (b) the kilowatthour, which is eqivalent to the energy supplied by a power of one thousand watts operating for one hour."

4.2.1.9 The unit of time interval is the Atomic Second, defined in 1967 by international agreement as a certain number of periods of a specified atomic transition of cesium 133.[5]

4.3 National Standards. The national standards of electrical measurement are those that are maintained by the National Bureau of Standards.

4.3.1 Standard of Resistance. The standard of resistance maintained by the National Bureau of Standards is based on absolute measurements in terms of length and time. The value of the national standard of resistance (the NBS ohm) is maintained in terms of the group average of a reference group of stable 1-ohm resistors whose values were initially assigned by international agreement based on the combined results of the various absolute-ohm determinations at the National Bureau of Standards and other national laboratories. The stability of these resistors is checked periodically by intercomparisons among the members of the group, by comparisons with the standards maintained by the national laboratories of other countries, and by a continuing series of absolute measurements.

---

[4]The power in an alternating-current circuit at any instant is the product of the current and the terminal voltage at that instant.

[5]The unit of time interval is not a part of Public Law 617, but has been accepted by the United States are the unit of time.

4.3.2 Standard of Electromotive Force. The standard of electromotive force maintained by the National Bureau of Standards is based on absolute measurements of the ohm and ampere in terms of mass, length, and time. The value of the national standard of EMF (the NBS volt) is maintained in terms of physical constants by means of the Josephson effect, which is used to check the stability of a reference group of saturated mercury-cadmium-amalgam standard cells.

4.3.3 Other Electrical Standards. The values of the national standards for other electrical quantities, such as energy, are derived either directly or indirectly from the values of the national standards of resistance and EMF, or (in the case of capacitance) obtained directly from computable standards constructed for absolute-ohm determinations.

4.3.4 Standard of Time Interval. Standard frequencies and time intervals are broadcast continuously by the National Bureau of Standards and the Naval Observatory. The time-interval signals as broadcast and received are of very high accuracy (much better than one part in a million), and are appropriate to use without corrections in verifying the rate of a laboratory standard clock or other reference interval timer.

4.4 Establishing a Local Reference Standard of Energy. Calibrations and tests whose purpose is the establishment or maintenance of a local reference standard of energy measurement (see 4.6.9) shall be carried out in a location and manner capable of maintaining the accuracy required of the standards used in verifying electricity meters. The sequential steps intermediate between the national standards of resistance and electromotive force and the local means of measuring energy are carried out at several functional levels, which may, but need not, be within the capabilities of a single laboratory. In many cases some or all of the steps may be carried out in the local meter laboratory. In others the local reference standard of energy may be compared (often by transport standards) with the national standard of electric energy or with standards of an independent laboratory that have been properly verified.

4.4.1 Meter Laboratory. The meter laboratory is concerned with:

(1) Maintaining standards whose values are assigned either directly or indirectly in terms of the national standards

(2) Assigning values to the working standards essential to the measurement of electrical quantities

It may be equipped and staffed to make calibration tests at some or all of the sequential steps intermediate between the national standards of resistance, EMF, and time interval, and a local reference standard of energy measurement (such as a group of watthour meters).

4.4.2 Meter Shop. The meter shop is concerned with the routine testing, repair, and calibration of electricity meters and of the auxiliary devices and equipment essential to the metering of electric energy and power.

4.4.3 Independent Standards Laboratory. An independent standards laboratory is maintained by and responsible to a company or authority other than the one that maintains the particular meter laboratory under consideration. If a meter laboratory does not carry out the entire chain of sequential measurements between the national electrical standards and the local energy standard, it must depend on an independent standards laboratory to make some of the required calibrations in this measurement chain. Alternatively, interlaboratory comparisons of reference standards with an independent standards laboratory are always informative and, in some instances, are needed to maintain the integrity of the local reference standards. In 4.6 the equipment required for a standards laboratory to make all of the sequential calibration steps of interest in a meter laboratory is listed. This listing may be of assistance in determining whether a particular standards laboratory is adequately equipped to perform a specific calibration task.

4.5 Laboratory Conditions. In a meter laboratory it is essential that ambient conditions, such as temperature and humidity, be maintained at values and within limits appropriate to the measurements made in the laboratory; and that other ambient factors that could interfere with proper measurement, such as atmospheric contaminations, mechanical disturbances, electrical and magnetic interference, and noise, be held to such levels that normal measurement techniques and results are not adversely affected.

4.5.1 Reference Temperature and Humidity. The ambient temperature in the meter laboratory shall be 23 °C,[6] with tolerances that depend upon the effects of temperature on the standards used and the apparatus under test. This temperature should be held constant not only during a test or calibration procedure, but also for a preceding period sufficient to ensure effective temperature equilibrium in the test equipment and in the device being tested. When the reference standards of a laboratory are assigned or checked at a temperature different from that at which they will be used, temperature corrections should be known and applied where significant.

The ambient relative humidity should be kept to values low enough that electrical insulation in the equipment used will not be affected. Relative humidities below 55% should be adequate for this purpose. In the absence of adequate shielding and guarding of laboratory instruments and circuits, the effects of bound electrostatic charges may be troublesome at very low humidities. However, shielding may well be a simpler and better solution to this problem than an attempt to hold humidity above some specified minimum value, say 40%. Any system that controls laboratory humidity within specified upper and lower limits should be designed to avoid excessive humidity in the event of failure of the control element.

---

[6]This temperature is specified in ANSI/IEEE C39.1-1981 [1].

4.5.2 Laboratory Power Sources. Direct- and alternating-current supplies used in the laboratory for calibration of instruments or meters, or for the measurement of voltage, current, or power, should be closely regulated, since fluctuations in the value being held can limit the accuracy of a calibration or measurement.

Rectified dc supplies should be substantially free from ripple, since the presence of ripple and its waveform have different effects on instruments having peak, average, or rms response.

Alternating-current supplies should be substantially free from waveform distortion, and the phase relation of combined current and voltage supplies should be capable of close regulation, since these factors may also influence calibration and measurement accuracy. For the most accurate watthour meter calibrations, the third harmonic in the current wave should not exceed 0.5% of the fundamental, and other harmonics in the current and voltage waves should not exceed 1.0%.

4.6 Laboratory Reference Standards. Laboratory reference standards are those standards that are used to assign and check the values of laboratory secondary standards.

4.6.1 Stability of Reference Standards. One of the most important characteristics of a reference standard is its stability; that is, the constancy of its assigned value with time. Hence, the use of reference standards should be limited to assigning and checking the values of secondary standards. Reference standards should not be exposed to the hazards of accidental misuse that occasionally occur in routine measurements. Further advantages may accrue if the basic reference standards of a laboratory never leave it; that is, are never subjected to transportation hazards. In this case, special transport standards must be available for the periodic interlaboratory comparison tests that act as a check on the stability of the basic reference standards.

4.6.2 Basic Reference Standards. The basic reference standards of a laboratory are those standards with which the values of the electrical units are maintained in the laboratory, and that serve as the starting point of the chain of sequential measurements carried out by the laboratory. If the laboratory is to perform the entire sequence of measurements between the national electrical standards and the local standard of energy measurement, its basic reference standards are standard cells (preferably saturated cells) and appropriate resistance standards.

4.6.2.1 Intercomparison. Ideally, the basic reference standards of a laboratory should be maintained in groups of three or more separate individual units that can be intercompared readily, since three is the minimum number of units for which a change in one of them can be both detected and located by intercomparison.

4.6.3 Transport Standards. Transport standards are standards of the same nominal value as the basic reference standards of a laboratory, and preferably of equal quality, which are regularly intercompared with the basic group but are reserved for the periodic interlaboratory comparison tests that act as checks on the stability of the basic reference group.

4.6.4 Standard Cells. The best reference standard of EMF is a group of saturated mercury-cadmium-amalgam standard cells maintained at a constant temperature in a stirred oil bath or in a constant-temperature air bath. In either case, adequately sensitive and stable means should be available to check the bath temperature and its constancy.

4.6.4.1 Unsaturated Standard Cells. A group of unsaturated cells, each of which has been allowed to come to equilibrium in a container designed to minimize temperature gradients, provides a standard that is generally adequate for meter laboratories, if the cells are intercompared frequently to guard against changes brought about by misuse or accident and if the EMFs of at least two of the cells have been checked against a saturated-cell reference standard within the preceding 12 months. (The EMF of an unsaturated cell generally decreases slowly with time, usually less than 100 microvolts a year.) The EMF of each cell in the local unsaturated group should be reassigned in terms of the transport cells checked against the saturated reference group, after they have been intercompared to ensure that they have not been damaged in transportation and have recovered stable values.

4.6.5 Standard Resistors. Minimum requirements should include reference standards at the 1-ohm level, and at decimal multiples and submultiples of the ohm, over the range of resistance required for both resistance and current measurement. Generally, standards covering the range from 10^{-4} to 10^4 ohms are useful, unless the range of measurements to be undertaken in the laboratory is specifically known to be less.

Additional standards having the following intermediate values of 0.002, 0.005, 0.02, 0.05, 0.2, 0.5, 2, 5, and 20 ohms are convenient since they permit the calibration of most values of current shunts by direct substitution techniques without the precise calibration of bridge ratio arms. Alternatively, a very precise 2:1 ratio can be established from the combinations of three nominally equal standards.

4.6.6 DC Ratio Devices. A dc ratio device is an arrangement of resistors for establishing one or more ratios in an accurately known way. Devices of this kind are used to compare the ratio of two resistors or of two direct voltages. An essential requirement of a dc reference ratio standard is that it be stable with time, and minimally affected by ambient conditions or by loading.

4.6.6.1 Reference Standard Volt Box. A reference standard volt box is one in terms of which the ratios of other volt boxes can be accurately assigned. It must have the following characteristics:

(1) It must be subdivided to facilitate internal self-calibration.

(2) Its components must be guarded or otherwise be adequate to eliminate errors from leakage currents across insulating members.

(3) It must be designed to avoid or minimize changes in ratio resulting from self-heating at rated voltage or from ambient temperature changes.

4.6.6.2 Direct-Reading Ratio Set. A direct-reading ratio set is a means for comparing nominally equal resistance standards. It should have a range of ±0.5% from equality (that is, ratios from 0.995 to 1.005).

4.6.6.3 Universal Ratio Set and Kelvin-Varley Divider. These are means for calibrating dc potentiometers and other adjustable resistance-type voltage dividers.

4.6.6.4 Thermofree Microvolt Potentiometer. A thermofree microvolt potentiometer is a means for intercomparing standard cells connected in opposition.

4.6.6.5 Resistance Bridges. The Wheatstone bridge for two-terminal resistance measurements and the Kelvin bridge for four-terminal measurements are indispensable intermediate-level calibration tools. Such bridges, of high quality and proven stability, can be used in substitution techniques for the intercomparison of reference standard resistors or for the assignment of working standard resistors of nominally equal values. For the general assignment of resistance values, where substitution techniques are not applicable, such bridges require periodic calibration, using standards of high quality in an appropriate technique.

4.6.6.6 Direct-Current Comparator. Appropriately designed direct-current comparators based on the equality of ampere-turns in two or more windings on a magnetic core are means for making dc ratio, resistance, and voltage measurements.

4.6.7 AC Ratio Devices. Current and voltage ratios are usually established in terms of instrument transformers. The ratio of the primary to secondary current (or voltage) and the phase angle between them are dependent on the magnitude of the primary quantity and on the burden of the secondary circuit. The ratio and phase-angle corrections of reference transformers shall be established at the expected operating level and for the burden with which the transformer will be used.

4.6.7.1 Stability. Since the corrections of an instrument transformer, for a given frequency and waveform, depend solely on the magnetic properties and condition of the core and on the geometry and resistance of the windings, these corrections should be extremely stable with time, provided that the core is demagnetized before use of the transformer as a reference standard.

4.6.7.2 Transformer Test Set. A transformer test set is a means of determining the corrections of an instrument transformer in terms of either an alternating-current comparator or a reference standard transformer, usually having the same nominal ratio.

4.6.8 DC-AC Transfer Standards. A transfer device is required to establish the equality of an rms alternating current or voltage, or the average value of alternating power, with the corresponding steady-state dc quantity that can be referred to the basic standards through potentiometric techniques. Such transfer standards for use at power frequencies employ electrothermic, electrodynamic, electrostatic, or electronic operating principles, although in US practice only the electrothermic and electrodynamic principles have been applied extensively. The electrodynamic principle of operation is generally used in power-transfer reference standards (electrodynamic watt-meters) in the United States.

4.6.8.1 Stability. The transfer characteristics of a transfer standard (that is, its ac-dc differences) are functions of its geometry, its electrical parameters, and its operating level, and should not change significantly with time. Hence, the transfer characteristics of an instrument need be determined only once, unless the components of its measuring circuit are modified or replaced, or their physical arrangement altered. However, the dc calibration of a transfer instrument should be verified periodically.

4.6.9 Reference Standard of Energy. One or more highly stable standard watthour meters operating under closely controlled conditions can be used to maintain the unit of energy in the laboratory and to calibrate the next level of standards. Such an energy standard is maintained in many laboratories as an essential part of their reference equipment, frequently in conjunction with instrument transformers having appropriate ratios, such that the reference standard watthour meter can always be operated at the same current and voltage level, regardless of the current or voltage requirements of the meters compared with the reference meters. Usually such standards are housed in a temperature-controlled environment, and are continuously energized. Alternatively, one or more electronic watthour meters of proven accuracy and stability may be used.

4.6.10 Time Interval. The reference standard of time interval in a laboratory may be a clock or other counting device whose rate is controlled by an appropriately compensated pendulum, or by a crystal or tuning-fork oscillator in a suitably controlled environment. Its operation should be such that the time-interval signals for laboratory use are produced without significant reaction on the rate-regulating mechanism. Such a standard can be checked at any time, or continuously monitored, using signals broadcast by the National Bureau of Standards or the Naval Observatory.

4.6.11 Periodic Verification of Reference Standards
4.6.11.1 Standard Cells. The reference standard cells with which the value of the volt is maintained in a laboratory should be intercompared at frequent intervals, and should be compared at suitable intervals with reference saturated cells maintained by an independent standards laboratory of recognized standing. If the local reference group consists of unsaturated cells, they should be compared with saturated reference cells at intervals not to exceed 1 year. If the reference group consists of saturated cells, the interval between comparisons with an independently maintained group of saturated

47

cells may be increased after the stability of the group has been established.[7]

4.6.11.2 Resistance Standards. The reference standard resistors, with which the resistance unit is maintained in a laboratory, should be intercompared frequently, and should be verified by an independent standards laboratory at intervals not longer than 2 years.

4.6.11.3 Ratio and Transfer Standards. Ratio standards (either ac or dc) and dc-ac transfer standards should be verified periodically or whenever there is reason to suspect a change in their performance, and when self-checking features do not eliminate the uncertainty in question.

4.6.11.4 Reference Standard of Energy. Watthour meters used as reference standards to maintain the unit of energy should be intercompared at frequent intervals. In addition, their registrations shall be determined periodically by appropriate calibrations.

4.7 Laboratory Secondary Standards. The secondary standards of a laboratory are those that are used in the routine calibration tasks of the laboratory. They are checked in terms of the laboratory reference standards, and are used for the calibration of shop instruments and meters, as well as for other routine measurement tasks for which moderately high accuracy is required.

4.7.1 Potentiometric Equipment. The basic working standard equipment of a meter laboratory should include one or more high-quality potentiometers with shunts and volt boxes having ranges appropriate for all required ammeter, voltmeter, and wattmeter calibration checks. An unsaturated standard cell or a source of reference voltage of equivalent quality (for example, a reference voltage regulated by a Zener diode in a suitably temperature-compensated network) is a necessary adjunct to a potentiometer.

4.7.2 Indicating Instruments. Analog (pointer and scale) or digital instruments of appropriate ranges and of high quality are required as laboratory secondary standards.

4.7.2.1 Accuracy Classes. It should be realized that the accuracy class designation (see ANSI/IEEE C39.1-1981 [1]) of an analog instrument is a specification of its performance under reference conditions and immediately after its circuit is energized. When the instrument is used under other than reference conditions, the stated class accuracy may not be realized; in particular, the influence of

---

[7]Many saturated cells are constructed without septa to hold the electrode materials in place. Such cells cannot be inverted and therefore cannot be shipped but must be hand-carried for interlaboratory comparisons. Some manufacturers place retaining septa in their saturated cells, and state that these cells can be shipped without damage. In any event, regardless of transport means, it is preferable to avoid either extremely hot or extremely cold weather, or any condition that would subject the cell to severe mechanical or thermal shock.

sustained operation on the instrument response may be significant. Hence, for best accuracy the conditions under which the dc calibration is checked should approximate the use condition as nearly as practicable. The difference between the ac response and the average of reversed dc indications for a signal of equivalent magnitude should not be significantly affected by time in circuit, so ac-dc differences need not be redetermined for different use situations.

The rated accuracy of a digital voltmeter conforming to ANSI C39.6-1969 (R 1975) [3], applies over a fairly wide range of operating conditions. Many ac digital voltmeters respond to the average rather than the rms value, and should be used with caution on distorted waveforms.

4.7.2.2 Instrument Bearings or Suspensions. Secondary-standard instruments having pivot-and-jewel bearings should have permanent locations in the laboratory, and should be moved as little as possible to avoid bearing damage with consequent increased friction. Instruments having taut-band suspensions are free from frictional errors and are not generally damaged by reasonable laboratory handling.

4.7.2.3 Range Extension. The range of ac instruments may be extended by the use of instrument transformers whose corrections are known for the particular instrument burden with which they are to be used. The range of ac instrument voltage circuits may also be extended by use of series resistors, but this means of range extension is usually limited to 600 V.

4.7.3 Watthour Meters. Secondary standard watthour meters of appropriate ranges, compensated as fully as practicable for the various known sources of error, are often required for determining the accuracy of registration of the portable standard watthour meters used in shop and field testing and adjustment operations.

4.7.4 Calibration Checks. Even when the secondary standards of a laboratory have been demonstrated to be stable, the hazards of accidental change or injury through misuse in the daily operation of the laboratory may be present. Hence, frequent calibration checks constitute insurance against faulty measurements remaining undetected for an extended period.

4.7.4.1 Potentiometer Accessories. Secondary standard resistors for current measurements (shunts), volt boxes, and standard cells should be verified at frequent and regular intervals in terms of the appropriate laboratory reference standard.

4.7.4.2 Potentiometer Calibrations. Potentiometer calibrations should be verified annually, either by using built-in self-calibration means (where provided), or alternatively with a universal ratio set. In addition, it is desirable that the calibration of a potentiometer be verified by an independent standards laboratory at intervals of not more than 3 years.

4.7.4.3 Indicating Instruments. One or more cardinal points of the dc calibration of secondary standard indicating instruments should be verified potentiometrically at intervals of 2 weeks to 3 months, depending on frequency of use; and a complete dc check throughout their working range should be made at intervals of 3 months to a year.

Direct- and alternating-current calibration consoles (adjustable, direct-reading power supplies) are very convenient for routine testing of instruments. The periodic calibrations of such consoles (usually with dc potentiometers and dc-ac transfer standards) should be carefully considered.

4.7.4.4 Watthour Meters. Watthour meters used as secondary standards should be intercompared at frequent and regular intervals, and their registration errors redetermined by a power-time integration procedure or by comparison with reference-standard watthour meters (see 4.6.9) at intervals not to exceed 6 months.

4.8 Shop Instruments. Shop instruments are instruments and meters that are used in regular routine shop or field operations. Their calibrations should be verified in terms of appropriate laboratory secondary standards.

4.8.1 Indicating Instruments. Portable ammeters, voltmeters, and wattmeters used in the regular operation of the meter shop should be of good quality and of appropriate ranges. Analog instruments should not be operated at less than one-third of their end-scale value. Depending on the accuracy desired for the particular measurement being made, Classes 0.25 or 0.5 (or in some instances even 1.0) may be appropriate. The corrections to shop instruments should be regularly and frequently redetermined, using laboratory secondary-standard instruments.

4.8.2 Portable Standard Watthour Meters. Portable standard watthour meters that are in constant use should be checked at least twice a month[8] on a commonly used current and voltage range. A complete check of all combinations should be made when the registration at the commonly used check points deviates by more than acceptable limits.

4.9 Performance Records. Continuing records should be kept of the performance of each instrument and standard in the laboratory or shop. Where this record shows excessive variation between tests, the equipment should be subjected to special investigation to determine the cause of the variation. If the cause cannot be determined and corrected, use of the instrument or standard should be discontinued.

4.9.1 Importance of Records. Continuing records on instruments and standards are important for a number of reasons:

(1) They can be very informative as to the quality of the laboratory equipment and the competence of laboratory personnel.

(2) The value of a standard increases as proof of its stability accumulates.

---

[8]Recommendations in Section 4 of this standard for the calibration of instruments, meters, and other measuring apparatus are based on average service in typical laboratories and shops. Longer intervals may be used if supported by adequate performance records.

(3) Continuous performance records assist in decisions whether to keep, demote, or discard a standard or instrument.

(4) Continuous records assist in decisions concerning the interval between calibration or verification tests of an instrument or standard.

4.10 Abnormal Conditions. Whenever a standard is suspected of having been subjected to abnormal conditions or treatment, it should be checked regardless of the time that has elapsed since its last calibration check. Resistance apparatus that has suffered an abrupt change in value because of misuse will sometimes drift during a few weeks or months before stabilizing at a different value.

4.11 Instrument Specification. ANSI C39.1-1981 [1], and ANSI C39.6-1969 (R1975) [3], are suggested as guides in specifying the quality and performance characteristics of portable and laboratory standard indicating instruments.

4.12 Acceptable Performance of Standard Watthour Meters
  4.12.1 General
    4.12.1.1 Acceptable Standard Watthour Meters. In order to be acceptable, new standard watthour meters shall be capable of conforming to mechanical requirements as specified in 4.12.2 and the performance requirements as specified in 4.12.3 and Table 4.12, which are intended to determine their reliability and acceptable accuracy insofar as these qualities can be demonstrated by laboratory tests. In general, all of the listed tests should be made on each new type of standard watthour meter and most of them should be made on each new standard meter of the same type, dependent on the needs of the particular laboratory. Two kinds of standard watthour meters are recognized: portable standard watthour meters, which are used for shop and field tests; and reference standard watthour meters, which are used to maintain the unit of energy in the meter laboratory.[9]

    4.12.1.2 Adequacy of Testing Laboratory. Tests for determining the acceptability of the types of standard meters under these specifications shall be made in a laboratory having adequate facilities, using instruments of an order of accuracy and precision capable of verifying conformance to the specifications. These instruments should be checked against the laboratory reference standards before and after the tests. The tests shall be conducted by personnel who have thorough practical and theoretical knowledge of meters and adequate training in making precision measurements.

    4.12.1.3 Tolerances. The standard meter under test shall be considered to be within the specified limit unless the test result exceeds the limit by more than the value of the measurement uncertainty

---

[9]For brevity, these are called portable standard meters and reference standard meters in 4.12. For requirements applicable to both types, the term *standard (watthour) meters* is used.

assigned to cover the possible errors in the laboratory reference standards, observations, and procedures.

4.12.2 Mechanical Requirements

4.12.2.1 General. All parts that are subject to corrosive influence under normal working conditions shall be effectively protected against corrosion due to atmospheric causes. Any protective coating shall not be liable to damage by ordinary handling or injuriously affected by exposure to air under ordinary conditions. The construction of the meter shall be suitable for its purpose in all respects, and shall give assurance of permanence in all mechanical, electrical, and magnetic adjustments.

4.12.2.2 Adjusting Devices. Adjusting devices shall be self-locking or, alternatively, shall be capable of being locked in position, and the action of such locking devices should not alter the adjustment of the standard meter. All mechanical, electrical, and magnetic adjustments shall be capable of fine control, and shall be of such design as will give assurance of permanence.

4.12.2.3 Leveling Means for Induction-Type Meters (Optional on Portable Standard Meters). A level indicator shall be provided on the top of the standard meter in a position easily read when the standard meter is in use. The sensitivity of the indicator shall be such that a departure from level of 0.5 degree is readily detectable. Facilities for adjusting the level of the standard meter shall be provided.

4.12.2.4 Case. The case shall be of sufficient strength to afford to the working parts adequate protection against damage under normal conditions of handling, usage, and transport; and it shall afford to the interior substantial protection against the entry of dust. Portable standard meters should be fitted with a detachable cover to enclose the readout, terminals, and controls, and be equipped with a substantial carrying strap. The inside of the cover should include a means for attaching a calibration card.

4.12.2.5 Sealing. Provision shall be made for the sealing of the standard meter to detect unauthorized access to working parts and to electrical and magnetic adjusting devices.

4.12.2.6 Window (Portable Standard Meters). A window of glass or other suitable transparent material shall be provided to permit a clear view of the readout. It shall be substantially dusttight and shall be replaceable.

4.12.2.7 Terminals. Terminal identification shall be adjacent to each terminal and shall be of a permanent nature.

4.12.2.8 Register (Counting Mechanism) (Portable Standard Meters)

4.12.2.8.1 Register Scales. For standard watthour meters with pointer and dial-type readouts, the register shall have a sweep-hand scale and not less than two totalizing scales. The tip of the sweep hand shall traverse its scale in such a manner as will permit accuracy of reading. One revolution of the sweep hand shall represent one revolution of the rotor. It shall be so designed as to minimize parallax error and be readable to 1/100 of a revolution. Totalizing scales shall be graduated and suitably marked in multiples of 10 and shall totalize to not less than 100 revolutions of the sweep hand.

4.12.2.8.2 Provision for Photoelectric Sensing. The sweep hand of standard watthour meters with pointer and dial-type readouts shall be provided with a reflective area suitable for external photoelectric sensing.

4.12.2.8.3 Digital Readouts. Standard watthour meters with a digital readout shall have a resolution corresponding to at least 1/1000 of a revolution of a pointer-type register for a meter of similar current and voltage range, and shall be capable of totalizing not less than 100 equivalent revolutions. Alternatively, they shall have a readout in other suitable units, such as percentage registration, with an equivalent resolution.

4.12.2.8.4 Readout Reset. A readily accessible reset device shall be fitted so that all pointers or the digits of the readout may be simultaneously reset to zero by a single operation. Such operation shall not permanently distort the spindles of the pointers (where applicable).

4.12.2.8.5 Register Lubrication. The register gear train shall not require lubrication.

4.12.2.9 Provision for Pulse Output. For reference standard meters, a pulse output shall be provided (optional on portable standards) such that the number of pulses is proportional to the energy measured. Connections to the pulse output shall be readily accessible on the outside of the meter.

4.12.2.10 Fuse (Portable Standard Meters). A suitable fuse, replaceable from the outside, shall be provided in any current coil circuit rated 1 ampere or less.

4.12.2.11 Nameplate. A nameplate shall be provided on the outside of the case to show all necessary information, including manufacturer, type, serial number, voltage ratings, current ratings, frequency, model number, and watthour constant (K_h) at basic voltage and current ratings.

4.12.2.12 Rotor Brake (Portable Standard Meters). A suitable rotor brake shall be provided to prevent rotor drift when current only is applied.

4.12.3 Performance Requirements

4.12.3.1 General Test Conditions. The standard meter under test shall be in good operating condition, and its registration on the basic current and voltage ranges shall be adjusted as nearly as practicable to 100% with 25% and 100% rated current at 1.0 power factor, and with 100% rated current at 0.5 power factor. The meter shall be energized on the basic range at 100% rated current and voltage for at least 1 hour prior to test. Unless otherwise specified, all tests shall be made on the basic range, and the conditions listed in 4.12.3.1.1 through 4.12.3.1.8 shall apply. All tests at other than unity power factor are with current lagging, unless otherwise noted.

4.12.3.1.1 Applied Voltage. The applied voltage shall be constant to within ± 1.0%.

4.12.3.1.2 Applied Current. The applied current shall be constant to within ± 1.0%.

4.12.3.1.3 Phase Angle. The phase angle shall be constant to within ± 2 degrees.

4.12.3.1.4 Frequency. The frequency shall be 60 Hz and be constant to within ± 0.2%.

4.12.3.1.5 Waveform Distortion. The total harmonic distortion of the applied voltage and current shall not exceed 2.0%.

4.12.3.1.6 Ambient Temperature. The ambient temperature shall be 23 °C ± 2 °C.

4.12.3.1.7 Level. Standard watthour meters of the induction type shall be level to within ± 0.5 degree.

4.12.3.1.8 External Magnetic Field. Strong magnetic fields may affect performance of standard meters. Care should be taken to avoid placing the meters in proximity to transformers and loops of current test leads.

4.12.3.2 Insulation. The insulation between current-carrying parts of separate circuits and between current-carrying parts and other metallic parts shall conform to ANSI C39.5-1974 [2], Section 11.5.5, and shall be capable of withstanding the application of a sinusoidal voltage of 2.3 kV rms, 60 Hz, for 1 minute.

4.12.3.3 Drift (Portable Standard Meters)

(1) With 250% rated current and with the voltage circuit open, the indication must not change perceptibly in 1 minute.

(2) With 100% rated current, the braking device on standard watthour meters of the induction type shall immediately release the rotor when 70% rated voltage is applied.

4.12.3.4 Effect of Variation of Current at 1.0 Power Factor

(1) With 50%, 150%, and 200% of rated current, the registration of a portable standard meter shall not differ from the value at 100% of rated current by more than the amount specified in Table 4.12.

(2) With 90% and 110% of rated current, the registration of a reference standard meter shall not differ from the value at 100% of rated current by more than the amount specified in Table 4.12.

4.12.3.5 Effect of Variation of Current at 0.5 Power Factor

(1) With 50% and 200% of rated current, the registration of a portable reference standard meter shall not differ from the value at 100% of rated current by more than the amount specified in Table 4.12.

(2) With 90% and 110% of rated current, the registration of a portable standard meter shall not differ from the value at 100% of rated current by more than the amount specified in Table 4.12.

4.12.3.6 Effect of Variation of Voltage At the Power Factors Indicated in Table 4.12

(1) With 25% of rated current at 1.0 power factor, the registration of a portable standard meter at 90% and 110% of rated voltage shall not differ from the value at 100% of rated voltage by more than the amount specified in Table 4.12.

(2) With 100% of rated current, the registration of a standard meter at 90% and 110% of rated voltage shall not differ from the

Table 4.12
Standard Watthour Meters

Section	Test Condition	% Maximum Deviation*	
		Portable Standard	Reference Standard
4.12.3.4 (Variation of current at pF 1.0)	(1) — (2) —	0.25 —	— 0.10
4.12.3.5 (Variation of current at pf 0.5)	(1) 50% *I* (1) 200% *I* (2) —	0.40 0.60 —	— — 0.10
4.12.3.6 (Variation of voltage)	(1) 1.0 pf (2) 1.0 pf (2) 0.5 pf	0.30 0.20 0.40	— 0.10 0.15
4.12.3.7.1 (Equality of current ranges)	1.0 pf 0.5 pf	0.20 0.40	0.10 0.10
4.12.3.7.2 (Equality of voltage range)	—	0.25	0.10
4.12.3.8 (Variation of temperature)	(1) 1.0 pf (1) 0.5 pf (2) 1.0 pf (2) 0.5 pf (3) 1.0 pf (3) 0.5 pf (4) 1.0 pf (4) 0.5 pf	0.30 0.50 0.30 0.50 — — — —	— — — — 0.10 0.15 0.10 0.15
4.12.3.9 (Effect of internal heating)	—	0.20	—
4.12.3.10 (Effect of tilt)	25% *I* 100% *I*	0.20 0.10	— —
4.12.3.11 (Repeatability)	—	0.20	0.05

*The % deviation denotes the maximum allowable difference between the percentage registrations noted at the reference condition and at the specified test condition.

value at 100% of rated voltage by more than the amount specified in Table 4.12.

NOTE: When a portable standard meter is furnished with an external multiplier for the purpose of extending the voltage range, this test shall include the extended voltage rating with the multiplier connected in the circuit.

55

4.12.3.7 Equality of Current and Voltage Ranges

4.12.3.7.1 Equality of Current Ranges. The standard shall be tested with rated current at 1.0 and 0.5 power factors for each current range with 100% of rated voltage. On all current ranges, the registration shall not differ from the registration on the basic current range by more than the amount specified in Table 4.12.

4.12.3.7.2 Equality of Voltage Ranges. The standard meter shall be tested with rated voltage at 1.0 and 0.5 power factors for each voltage range with 100% of rated current on the basic current range. On all voltage ranges, the registration shall not differ from the registration on the basic voltage range by more than the amount specified in Table 4.12.

When a portable standard meter is furnished with an external multiplier for the purpose of extending the voltage range, this test shall include the extended voltage rating with the multiplier connected in the circuit.

4.12.3.8 Effect of Variation of Ambient Temperature. The test conditions are as follows: The standard meter shall be placed in a space having a temperature of 23 °C ± 2 °C and allowed to stand for not less than 2 hours with the voltage circuit energized. The meter shall then be tested with 100% of rated current at 1.0 and 0.5 power factors.

Conditions (1) and (2), following, apply to portable standard meters. Conditions (3) and (4) apply to reference standard meters.

Condition (1). The portable standard meter shall be placed in a space having a temperature of 0 °C ± 5 °C for not less than 2 hours with the voltage circuit energized. The meter shall then be tested with 100% of rated current at 1.0 and 0.5 power factors.

Condition (2). Repeat condition (1), except that the portable standard meter shall be placed in a space having a temperature of 50 °C ± 5 °C.

Condition (3). The reference standard meter shall be placed in a space having a temperature of 11 °C ± 2 °C for not less than 2 hours with the voltage circuit energized. The meter shall then be tested with 100% of rated current at 1.0 and 0.5 power factors.

Condition (4). Repeat condition (3), except that the reference standard meter shall be placed in a space having a temperature of 35 °C ± 2 °C.

At 0 °C ± 5 °C and 50 °C ± 5 °C the registration of a portable standard meter shall not differ from the value at 23 °C ± 2 °C by more than the amount specified in Table 4.12.

At 11 °C ± 2 °C and 35 °C ± 2 °C the registration of a reference standard meter shall not differ from the value at 23 °C ± 2 °C by more than the amount specified in Table 4.12.

4.12.3.9 Effect of Internal Heating (Not Applicable for Reference Standard Meters). The basic range of the portable standard meter shall be tested with 100% of rated current at 1.0 and 0.5 power factors. Then, 200% rated current shall be applied to the highest current range for 2 hours and the basic range shall be re-tested immediately afterward with 100% of rated current at 1.0 and 0.5 power factors. The change in registration shall not exceed the amount specified in Table 4.12.

4.12.3.10 Effect of Tilt (Not Applicable for Reference Standard Meters). The basic range of the portable standard meter shall be tested with 25% and 100% of rated current at 1.0 power factor. The meter shall then be tilted 1 degree forward, backward, to the left, and to the right, and tested in each position with 25% and 100% of rated current at 1.0 power factor. The change in registration of any of these positions shall not exceed the amount specified in Table 4.12.

4.12.3.11 Repeatability of Performance. The standard meter shall be placed in a space having a temperature of 23 °C ± 2 °C, and operated continuously with 10% of rated current at 1.0 power factor for a period of 1 week. The percentage registration, at rated voltage, at 25% and 100% of rated current shall be determined at the start of the test and at four successive intervals, at least 24 hours apart, during the week. The change in percent registration shall not differ from that at the start of the test by more than the amount specified in Table 4.12.

5. Acceptable Performance of New Types of Electricity Meters and Instrument Transformers

5.1 Watthour Meters
5.1.1 General
5.1.1.1 Acceptable Meters. New types of electro-mechanical meters, in order to be acceptable, shall conform to certain requirements specified in 5.1.3 through 5.1.8, which are intended to determine their reliability and acceptable accuracy insofar as these qualities can be demonstrated by laboratory tests.

5.1.1.2 Adequacy of Testing Laboratory. Tests for determining the acceptability of the types of meters under these specifications shall be made in a meter laboratory having adequate facilities, using instruments of an order of accuracy at least equal to that of the shop instruments and standards described in Section 4, Standards and Standardizing Equipment. These instruments should be checked against the laboratory secondary standards before and after the tests, or more often as required. The tests shall be conducted only by personnel who have thorough practical and theoretical knowledge of meters and adequate training in the making of precision measurements.

5.1.2 Definitions. The following terms are used in this section and are defined in Section 2, Definitions:
  (1) approval test (see 2.89)
  (2) class designation (see 2.109)
  (3) creep (see 2.110)
  (4) form designation (see 2.111)
  (5) load range (see 2.117)
  (6) meter laboratory (see 2.42)
  (7) percentage error of a meter (see 2.119)
  (8) percentage registration of a meter (see 2.120)
  (9) reference performance (see 2.124)
  (10) test current (see 2.133)
  (11) two-rate meters (see 2.134)
  (12) waveform distortion-percent (see 2.136)

5.1.3 Types of Watthour Meters
5.1.3.1 Basic Type. Meters are considered to be of the same basic type if they are produced by the same manufacturer, bear a related type designation, are of the same general design, and have the same relationship of parts. They must be substantially equivalent in the following respects:
  (1) Arrangement and shape of magnetic circuits
  (2) Electric circuits and connections
  (3) Heavy-load torque
  (4) Heavy-load speed
  (5) Design of retarding-magnet system

(6) Design and weight of rotor
(7) Design and type of rotor bearings

5.1.3.2 Variations within the Basic Type. Meters of the same basic type may vary according to the service for which they are designed, namely:
(1) Voltage rating
(2) Class and test-ampere rating
(3) Frequency
(4) Two-, three-, four-, or five-wire service
(5) Single stator or multistator
(6) Wye or delta service
(7) Method of mounting: detachable (Type S) or provided with terminal compartment (Type A), or switchboard mounting

NOTE: The factors given in 5.1.3.2(4) through 5.1.3.2(7) are interrelated with the watthour meter form designation (see 2.111).

5.1.3.3 Type Designation. Meters of the same basic type, but differing in the number of stators, shall be assigned type designations that will identify both the basic type and number of stators. In addition, two-stator three-wire meters, two-stator four-wire delta meters, and two-stator four-wire wye meters shall be assigned different type designations.

5.1.3.4 Acceptance of Basic Types in Whole or Part. A basic type of meter may be accepted as a whole, or a restricted variation of a basic type may be accepted.

5.1.3.5 Minor Variations. Minor variations in the mechanical construction, which are not of such character as to affect the electrical operation of the meter, may be permitted in the different meters of the same basic type.

5.1.3.6 Meters Requiring Separate Tests. Meters of the same basic type, but differing in frequency, shall be treated as different types for purposes of approval tests.

Two-wire and three-wire single-stator meters of the same type are to be treated as different types for purposes of approval tests, unless they are in fact similar in all essential details.

Single-stator and multi-stator meters of the same basic type are to be treated for purposes of acceptance tests as if they were of different types.

In the case of meters designed and adjusted to be used with specific current transformers, or current and voltage transformers, these specifications apply to the performance of the meters only.

5.1.3.7 Special Types. In the case of a type of meter that comes within the scope of these specifications, but is of such design that some of the tests hereinafter specified are inapplicable or cannot be made under the specified conditions, limited approval may be granted subject to appropriate restrictions.

5.1.4 Specifications for Design and Construction

5.1.4.1 Type Designation and Identification. Each meter shall be designated by type and given a serial number by the manufacturer.

The serial number and type designation shall be legibly marked on the nameplate of each meter. The register ratio shall be marked on a permanent part of the register.

5.1.4.2 Sealing. Meters shall be provided with facilities for sealing to prevent unauthorized entry.

5.1.4.3 Cover. The cover shall be dustproof, and shall be raintight if intended for outdoor installation.

5.1.4.4 Terminals. The terminals of the meter shall be so arranged that the possibility of short circuits is minimized when the cover is removed or replaced, connections are made, or the meter is adjusted.

5.1.4.5 Construction and Workmanship. Meters shall be substantially constructed of good material in a workmanlike manner, with the objective of attaining stability of performance and sustained accuracy over long periods of time and over wide ranges of operating conditions with a minimum of maintenance.

5.1.4.6 Weight of Rotor. The rotor shall be as light as practicable without the sacrifice of other desirable features.

5.1.4.7 Rated-Test-Current Torque-to-Friction Ratio. The ratio of the rated-test-current torque to the friction at the speed corresponding to rated test current shall be high, thus minimizing the possibility of change in the accuracy of the meter due to changes in friction.

5.1.4.8 Fixed and Adjustable Parts. All fixed parts shall be held securely in a permanent relationship. All adjustable parts shall be so constructed that they can be readily released, easily moved, and securely fastened in place without damage to the parts or to the meter.

5.1.4.9 Provision for Adjustment. Connections and parts requiring adjustment in service shall be easily accessible after the cover is removed. The adjustments shall permit calibration of the meter at 10% and 100% of nameplate test current (TA), 1.0 power factor, to provide correct registration under all ordinary conditions met in service, including the use of auxiliary apparatus.

5.1.5 Selection of Meters for Approval Tests

5.1.5.1 Samples to be Representative of the Basic Type. The meters to be tested shall be representative of the basic type and shall represent the average commercial product of the manufacturer.

5.1.5.2 Number to be Tested. A minimum of eight meters shall be subjected to test, except in the case of multi-stator meters and meters of an unusual or little-used type when a smaller number may be taken as being representative of specific type designations.

When the samples representing a given basic type include:

(1) Different current ratings, there shall be not less than two identical meters of each of the representative current ratings for each group

(2) Different voltage ratings, there shall be not less than two identical meters of each of the representative voltage ratings for each group

(3) Different number of stators, there shall be not less than two

identical meters of each representative number of stators for each group.

For the purpose of this section, four-wire wye and delta meters of the two-stator type are to be considered as separate categories.

When the test of a basic meter type includes Class 200 meters, at least 25% of all meters tested should be Class 200.

5.1.5.3 Additional Meters for Replacements. When practicable, meters submitted for approval tests should be accompanied by a sufficient additional number of each variation within the basic type, from which meters found defective or those accidentally damaged may be replaced.

5.1.6 Conditions of Test

5.1.6.1 Tests to be Applied to All Meters. Each meter shall be subjected to the tests as specified in 5.1.8, except meters may be exempted from certain tests, if they are meters for special services or if they are a modification of a type that has already been subjected to the test.

When meters are used as replacements for those meters that are found to fail some tests, be defective, or be damaged during test, such replacement meters shall be subjected to all of the required tests.

5.1.6.2 Waveform Distortion. All alternating-current tests shall be conducted on a circuit supplied by a sine-wave source with a distortion factor not greater than 3%.

5.1.6.3 Covers and Registers. Meter covers, where possible, shall be in place during all tests.

The registers of the meters shall be properly meshed during the tests.

5.1.6.4 Order of Conducting Tests. The items of each test shall be conducted in the order given.

After each change in voltage or load, a sufficient time interval shall be allowed for the meter to come to a stable condition before making the next observation or test.

5.1.6.5 Meters for Special Services. Meters designed for specific types of services may be tested on the type of circuit for which the meters are designed, or by using single-phase circuits provided that the meters meet the requirements of the test for independence of stators. In such cases, the testing laboratory may modify the procedures outlined herein to meet the requirements for single-phase tests on such types of meters. Examples are as follows:

(1) For volt-square-hour meters or stators, the *test current in amperes* as given in the specifications, shall not apply. These meters or stators shall be tested at 80%, 100%, and 120% of nameplate voltage.

(2) Meters used to measure quantities, such as varhours or *Q*-hours, may be tested in accordance with the test specifications for watthour meters, insofar as the tests apply.

(3) Where single-stator 240 V three-wire meters are to be approved for use on 120 V two-wire service, they shall be designed to meet all

Table 5.1.8.3
Creep Test

Class	Calibration Current (Amperes)
10	0.25
20	0.25
100	1.50
200	3.00
320	5.00

the applicable test requirements of this standard at both voltages, except that in the test for creep (see 5.1.8.3) at 120 V, the meter shall be adjusted to be 1.5% fast at the calibration current specified in Table 5.1.8.3. In the test for effect of tilt (see 5.1.8.17) the allowable maximum percent deviation from reference performance shall be ± 2.0% for conditions (1), (2), (3), and (4) on the 120 V test.

(4) For meters for special purposes or measurements, acceptance tests may be made on types not otherwise covered herein.

5.1.6.6 Meters of Non-Standard Classes. When meters of a class other than 10, 20, 100, 200, and 320 are submitted for acceptance, the *test current in amperes* for all tests under the sections on performance requirements shall be as recommended by the manufacturer.

5.1.7 Rules Governing the Acceptance of Types

5.1.7.1 Replacements. Replacements or repairs may be made if physical defects of a minor nature become apparent during the tests. If, during the test of a watthour meter, significant defects in design or manufacture become apparent, the test shall be suspended.

5.1.7.2 Tolerances. Due to possible errors in observations and in the standards employed, a tolerance should be applied to the specified limits of percent deviation for any test condition involving a determination of the accuracy of the meter. A meter should not be considered outside of the allowable deviation from reference performance unless the deviation is exceeded by 0.1% or by one-tenth of the allowable deviation, whichever is greater.

5.1.7.3 Basis of Acceptable Performance. A type of meter shall be acceptable under these specifications when the following requirements are satisfied:

(1) In 5.1.8.2 through 5.1.8.20, in the tests made on eight meters, not over three meters shall fail in a given item of a given subsection.

(2) In 5.1.8.2 through 5.1.8.20, an individual meter shall not fail in more than five items.

(3) In 5.1.8.2 through 5.1.8.20, in the tests made on eight meters, the total number of failures, including all the meters and all the items of each subsection, shall not exceed sixteen.

(4) In 5.1.8.2 through 5.1.8.20, in any of the subsections in which tests on fewer than eight meters are permitted, a meter shall not fail in any item in such tests. When the results of tests made on less than

Table 5.1.8.4
Starting Load Test

Class	Current (Amperes)
10	0.025
20	0.025
100	0.15
200	0.30
320	0.50

eight meters are considered unreliable, the tests may be made on eight meters, and the type adjudged in accordance with the requirements of 5.1.7.3(1) through 5.1.7.3(3).

5.1.8 Performance Requirements

5.1.8.1 Initial Conditions. The meter shall be in good operating condition, and it shall be adjusted as nearly as practicable to 100% registration at 10% and 100% of its nameplate test current value (TA) and at nameplate voltage. Where the meter has more than one voltage or current circuit, it shall be adjusted with the voltage circuits in parallel and all current circuits in series. In addition, for a multistator meter the stators shall be adjusted (balanced) to have performance equal to each other at 1.0 power factor 100% nameplate test current (TA). Each stator performance shall also be adjusted at 0.5 power factor lag 100% nameplate test current (TA) to its performance at 1.0 power factor.

Unless otherwise stated, all performance tests shall be made with the meter connected in the same manner as for adjustment. Where specific values are indicated, the meter shall be adjusted and operated as nearly as practicable to those values.

5.1.8.2 Insulation. The insulation between current-carrying parts of separate circuits and between current-carrying parts and other metallic parts shall be capable of withstanding the application of a sinusoidal voltage of 2.5 kV rms, 60 Hz, for 1 minute.

5.1.8.3 Test No 1: Creep. The test shall be made at rated frequency at an ambient temperature of 23 °C ± 5 °C and 100% of nameplate voltage.

The rotor of a meter shall not make one complete revolution in the forward direction at no load when the meter is adjusted at the calibration current specified in Table 5.1.8.3 to be 2% fast relative to registration at 100% nameplate test current (TA).

5.1.8.4 Test No 2: Starting Load. The test shall be made at nameplate voltage, rated frequency, 1.0 power factor, and at an ambient temperature of 23 °C ± 5 °C.

The rotor of a meter shall rotate continuously with a load current as specified in Table 5.1.8.3.

Table 5.1.8.5
Load Performance Test

| Condition | Meter Class | | | | | Maximum Deviation in Percent from 100% Registration |
| | 10 | 20 | 100 | 200 | 320 | |
			(Current in Amperes)			
(1)	0.15	0.15	1	2	3	± 2.0
(2)	0.25	0.25	1.5	3	5	—
(3)	0.5	0.5	3	6	10	± 1.0
(4)	1.5	1.5	10	20	30	± 1.0
(5)	2.5	2.5	15	30	50	—
(6)	—	5	30	60	75	± 1.0
(7)	5	10	50	100	100	± 1.5
(8)	7.5	15	75	150	150	± 2.0
(9)	—	18	90	180	250	± 2.0
(10)	10	—	100	200	300	± 2.0
(11)	—	20	—	—	320	± 2.5

5.1.8.5 Test No 3: Load Performance. The test shall be made at nameplate voltage, rated frequency, 1.0 power factor, and at an ambient temperature of 23 °C ± 5 °C.

The performance of the meter shall not deviate from 100% registration by an amount exceeding that specified in Table 5.1.8.5, except that the tests for conditions (9) through (11) shall be omitted for two-stator four-wire-wye meters.

Table 5.1.8.6.1
Effect of Variation of Power Factor
for Single-Stator Meters

Condition	Meter Class				Power Factor	Maximum Deviation in Percent from Reference Performance
	10	100	200	320		
	(Current in Amperes)					
Reference performance for condition (1)	0.25	1.5	3	5	1.0	—
Condition (1)	0.5	3	6	10	0.5 lag	± 2.0
Reference performance for condition (2)	5	50	100	150	1.0	—
Condition (2)	5	50	100	150	0.5 lag	± 2.0
Reference performance for condition (3)	10	100	200	320	1.0	—
Condition (3)	10	100	200	320	0.5 lag	± 2.0

5.1.8.6 Test No 4: Effect of Variation of Power Factor. The test shall be made at nameplate voltage, rated frequency, and at an ambient temperature of 23 °C ± 5 °C. Each stator of a multistator meter shall be tested as a single-stator meter, except that all voltage circuits shall be in parallel. The specified power factor refers to the power-factor relation of the current and voltage applied to each meter.

5.1.8.6.1 Single-Stator Meters. The effect of variation of power factor upon the performance of the meter shall not exceed that specified in Table 5.1.8.6.1.

Table 5.1.8.6.2
Effect of Variation of Power Factor for Two-Stator Meters: Network, Three-Phase Three-Wire, Three-Phase Four-Wire Delta, and Two-Phase Five-Wire

Condition	Meter Class 10	20	100	200 (Current in Amperes)	Power Factor	Maximum Deviation in Percent from Reference Performance
Reference performance for conditions (1) and (2)	0.5	0.5	3	6	1.0	—
Condition (1)	0.5	0.5	3	6	0.866 lead	± 2.0
Condition (2)	1.0	1.0	6	12	0.5 lag	± 2.0
Reference performance for condition (3)	2.5	2.5	15	30	1.0	—
Condition (3)	2.5	2.5	15	30	0.866 lead	± 1.0
Reference performance for conditions (4) and (5)	5	10	50	100	1.0	—
Condition (4)	5	10	50	100	0.866 lead	± 1.0
Condition (5)	5	10	50	100	0.5 lag	± 1.5
Reference performance for conditions (6) and (7)	10	20	100	200	1.0	—
Condition (6)	10	20	100	200	0.866 lead	± 1.5
Condition (7)	10	20	100	200	0.5 lag	± 2.0

NOTE: For Class 320, 2-stator 3-wire network and delta meters, the deviation limits given in Table 5.1.8.6.2 are recommended for the following test currents:
Reference for conditions (1) and (2): 10 A
Test condition (1): 10A, condition (2): 20 A
Reference for condition (3): 50 A
Test condition (3): 50 A
Reference for conditions (4) and (5): 150 A
Test conditions (4) and (5): 150 A
Reference for conditions (6) and (7): 320 A
Test conditions (6) and (7): 320 A

5.1.8.6.2 Two-Stator Network Meters, Two-Stator Three-Phase Three-Wire Meters, Two-Stator Three-Phase Four-Wire Delta Meters, and Two-Stator Two-Phase Five-Wire Meters. The effect of variation of power factor upon the performance of the meter shall not exceed that specified in Table 5.1.8.6.2.

Table 5.1.8.6.3
Effect of Variation of Power Factor
for Two-Stator Three-Phase Four-Wire-Wye Meters

Condition	Meter Class				Power Factor	Maximum Deviation in Percent from Reference Performance
	10	20	100	200		
	(Current in Amperes)					
Reference performance for conditions (1) and (2)	1	1	6	12	1.0	—
Condition (1)	1	1	6	12	0.866 lead	± 2.0
Condition (2)	2	2	12	24	0.5 lag	± 2.0
Reference performance for condition (3)	5	10	30	60	1.0	—
Condition (3)	5	10	30	60	0.866 lead	± 1.0
Reference performance for conditions (4) and (5)	10	20	100	200	1.0	—
Condition (4)	10	20	100	200	0.866 lead	± 1.0
Condition (5)	10	20	100	200	0.5 lag	± 1.5

5.1.8.6.3 Two-Stator Three-Phase Four-Wire-Wye Meters. The effect of variation of power factor upon the performance of the meter shall not exceed that specified in Table 5.1.8.6.3. Load current shall not be applied to the current winding that is common to both stators.

Table 5.1.8.6.4
Effect of Variation of Power Factor
for Three-Stator Three-Phase Four-Wire-Wye Meters

	Meter Class					Maximum Deviation in Percent from Reference Performance
Condition	10	20	100 (Current in Amperes)	200	Power Factor	
Reference performance for condition (1)	0.5	0.5	3	6	1.0	—
Condition (1)	1	1	6	12	0.5 lag	± 2.0
Reference performance for condition (2)	5	10	50	100	1.0	—
Condition (2)	5	10	50	100	0.5 lag	± 1.5
Reference performance for condition (3)	10	20	100	200	1.0	—
Condition (3)	10	20	100	200	0.5 lag	± 2.0

NOTE: For Class 320 meters, the deviation limits given in Table 5.1.8.6.4 are recommended for the following test currents:
  Reference for condition (1): 10 A
  Test condition (1): 20 A
  Reference and test condition (2): 150 A
  Reference and test condition (3): 320 A

  5.1.8.6.4 Three-Stator Three-Phase Four-Wire-Wye Meters. The effect of variation of power factor upon the performance of the meter shall not exceed that specified in Table 5.1.8.6.4.

Table 5.1.8.7
Effect of Variation of Voltage

| Condition | Meter Class | | | | | Maximum Deviation in Percent from Reference Performance |
| | 10 | 20 | 100 | 200 | 320 | |
	(Current in Amperes)					
Reference performance 100% of calibration voltage for conditions (1) and (2)	0.25	0.25	1.5	3	5	—
Condition (1) 90% of calibration voltage	0.25	0.25	1.5	3	5	± 1.0
Condition (2) 110% of calibration voltage	0.25	0.25	1.5	3	5	± 1.0
Reference performance 100% of calibration voltage for conditions (3) and (4)	2.5	2.5	15	30	50	—
Condition (3) 90% of calibration voltage	2.5	2.5	15	30	50	± 1.0
Condition (4) 110% of calibration voltage	2.5	2.5	15	30	50	± 1.0

5.1.8.7 Test No 5: Effect of Variation of Voltage. The test shall be made at rated frequency, 1.0 power factor, and at an ambient temperature of 23 °C ± 5 °C. The effect of variation of voltage upon the performance of a meter shall not exceed that specified in Table 5.1.8.7.

Table 5.1.8.8
Effect of Variation of Frequency

Condition	Meter Class 10	20	100	200	320	Percent Rated Frequency	Maximum Deviation in Percent from Reference Performance
	(Current in Amperes)						
Reference per-formance for conditions							
(1) and (2)	0.25	0.25	1.5	3	5	100	—
Condition (1)	0.25	0.25	1.5	3	5	95	± 1.0
Condition (2)	0.25	0.25	1.5	3	5	105	± 1.0
Reference per-formance for conditions							
(3) and (4)	2.5	2.5	15	30	50	100	—
Condition (3)	2.5	2.5	15	30	50	95	± 1.0
Condition (4)	2.5	2.5	15	30	50	105	± 1.0

5.1.8.8 Test No 6: Effect of Variation of Frequency.[10] The test shall be made at nameplate voltage, 1.0 power factor, and at an ambient temperature of 23 °C ± 5 °C. The effect of variation of frequency upon the registration of a meter carrying constant load shall not exceed that specified in Table 5.1.8.8.

---

[10] This test may be omitted where certified data are furnished by the manufacturer.

Table 5.1.8.9.1
Equality of Current Circuits for Single-Stator Meters

Condition	Connections of Current Circuits	Meter Class				Maximum Deviation in Percent from Reference Performance
		10	100	200	320	
		(Current in Amperes)				
Reference performance for conditions (1) and (2)	Both circuits	0.25	1.5	3	5	—
Condition (1)	Circuit A only	0.5	3	6	10	± 1.0
Condition (2)	Circuit B only	0.5	3	6	10	± 1.0
Reference performance for conditions (3) and (4)	Both circuits	2.5	15	30	50	—
Condition (3)	Circuit A only	5	30	60	100	± 1.0
Condition (4)	Circuit B only	5	30	60	100	± 1.0

5.1.8.9 Test No 7: Equality of Current Circuit. The test shall be made at nameplate voltage, rated frequency, 1.0 power factor, and at an ambient temperature of 23 °C ± 5 °C.

5.1.8.9.1 Single-Stator Meters. The change produced in the performance of a three-wire meter by using only one current circuit as compared with that when both current circuits are used shall not exceed that specified in Table 5.1.8.9.1.

Table 5.1.8.9.2 (1)
Equality of Current Circuits for Multistator Meters
Having One or More Three-Wire Stators

Condition	Connections of Current Circuits	Meter Class				Maximum Deviation in Percent from Reference Performance
		10	20	100	200	
		(Current in Amperes)				
Reference performance for conditions (1) and (2)	Both circuits	0.5	0.5	3	6	—
Condition (1)	Circuit A only	1	1	6	12	± 1.0
Condition (2)	Circuit B only	1	1	6	12	± 1.0
Reference performance for conditions (3) and (4)	Both circuits	5	5	30	60	—
Condition (3)	Circuit A only	10	10	60	120	± 1.0
Condition (4)	Circuit B only	10	10	60	120	± 1.0

5.1.8.9.2 Multistator Meters

(1) The change produced in the performance of a multistator meter having one or more three-wire stators by using only one current circuit of the three-wire stator as compared with that when both current circuits of the stator are used, shall not exceed that specified in Table 5.1.8.9.2 (1). These tests shall be made on each stator separately with no current flowing in the current circuits of the other stator or stators, but with the voltage circuits of all stators energized in parallel at rated voltage.

(2) The change produced in the performance of a multistator meter by using only one current circuit as compared with that when all current circuits are used, shall not exceed that specified in Table 5.1.8.9.2 (2). The current circuits that are not common to both stators of a two-stator three-phase four-wire wye meter shall be loaded at twice the test current specified. The circuits of any three-wire stator shall be connected in series and treated as one circuit.

Table 5.1.8.9.2 (2)
Equality of Current Circuits for Multistator Meters

Condition	Connections of Current Circuits	Meter Class (Current in Amperes)				Maximum Deviation in Percent from Reference Performance
		10	20	100	200	
Reference performance for conditions (5), (6), (7), (8), etc	All circuits	0.25	0.25	1.5	3	—
Condition (5)	Circuit A only	0.25 N*	0.25 N*	1.5 N*	3 N*	± 1.5
Condition (6)	Circuit B only	0.25 N*	0.25 N*	1.5 N*	3 N*	± 1.5
Conditions (7), (8), etc	Circuits C, D, etc	0.25 N*	0.25 N*	1.5 N*	3 N*	± 1.5
Reference performance for conditions (9), (10), (11), (12), etc	All circuits	2.5	2.5	15	30	—
Condition (9)	Circuit A only	2.5	2.5	15	30	± 1.5
Condition (10)	Circuit B only	2.5	2.5	15	30	± 1.5
Conditions (11), (12), etc	Circuits C, D, etc	2.5	2.5	15	30	± 1.5

* N represents the number of stators in the meter.

NOTE: For Class 320 meters, the deviation limits given in Table 5.1.8.9.2 (2) are recommended for the following test currents:

Reference for conditions (5), (6), (7), (8), etc: 5 A
Test conditions (5), (6), (7), (8), etc: 5 N* A
Reference and test conditions (9), (10), (11), (12), etc: 50 A

5.1.8.10 Test No 8: Effect of External Magnetic Field. The test shall be made at nameplate voltage, rated frequency, and 1.0 power factor at an ambient temperature of 23 °C ± 5 °C and shall be applied to one meter of each class.

The external alternating magnetic field, of the same frequency as that of the testing current, shall be produced by a straight conductor 6 feet long with return leads arranged to form a 6-foot square. A current in phase with the voltage applied to the meter shall be passed through this conductor. The return leads of the conductor shall be so arranged that the loop that they form does not surround or include the meter. The straight 6-foot conductor shall be placed in each of the following positions:

Condition (1). Behind the test board in a horizontal position and parallel to the back of the meter. The middle of the conductor shall be 10 inches directly behind and on a level with the center of the rotor. The loop shall be in a horizontal plane perpendicular to the test board.

Condition (2). Directly behind the centerline of the meter in a vertical position. The middle of the conductor shall be 10 inches directly behind and on a level with the center of the rotor. The loop shall be in a vertical plane perpendicular to the test board.

Condition (3). Vertically at the same distance in front of the test board as the center of the rotor. The middle of the conductor shall be on a level with the center of the rotor and 10 inches to the right or left. The loop shall be in vertical plane parallel to the test board.

For conditions (1) through (3) the change produced in the performance of a meter by the application of a 100-ampere-turn external magnetic field shall not exceed that specified in Table 5.1.8.10.

Table 5.1.8.10

Effect of External Magnetic Field

Condition	Meter Class (Current in Amperes)					Position of Conductor	Maximum Deviation in Percent from Reference Performance
	10	20	100	200	320		
Reference performance for conditions (1), (2), and (3)	0.25	0.25	1.5	3	5	—	—
Condition (1)	0.25	0.25	1.5	3	5	Condition (1) in 5.1.8.10	± 1.0
Condition (2)	0.25	0.25	1.5	3	5	Condition (2) in 5.1.8.10	± 1.0
Condition (3)	0.25	0.25	1.5	3	5	Condition (3) in 5.1.8.10	± 1.0

Table 5.1.8.11
Effect of Variation of Ambient Temperature

Condition	Meter Class (Current in Amperes)					Power Factor	Ambient Temperature	Maximum Deviation in Percent from Reference Performance at Nominal Temperature Difference*
	10	20	100	200	320			
Reference performance for conditions (1) and (7)	0.25	0.25	1.5	3	5	1.0	23 °C ± 5 °C	—
Reference performance for conditions (2) and (8)	2.5	2.5	15	30	50	1.0	23 °C ± 5 °C	—
Reference performance for conditions (3) and (9)	5	10	50	100	150	1.0	23 °C ± 5 °C	—
Reference performance for conditions (4) and (10)	0.5	0.5	3	6	10	0.5 lag	23 °C ± 5 °C	—
Reference performance for conditions (5) and (11)	2.5	2.5	15	30	50	0.5 lag	23 °C ± 5 °C	—

(Continued on Page 77)

Table 5.1.8.11 (Continued)
Effect of Variation of Ambient Temperature

Condition	Meter Class (Current in Amperes) 10	20	100	200	320	Power Factor	Ambient Temperature	Maximum Deviation in Percent from Reference Performance at Nominal Temperature Difference*
Reference performance for conditions (6) and (12)	5	10	50	100	150	0.5 lag	23 °C ± 5 °C	—
Condition (1)	0.25	0.25	1.5	3	5	1.0	50 °C ± 5 °C	± 2.0
Condition (2)	2.5	2.5	15	30	50	1.0	50 °C ± 5 °C	± 1.0
Condition (3)	5	10	50	100	150	1.0	50 °C ± 5 °C	± 1.0
Condition (4)	0.5	0.5	3	6	10	0.5 lag	50 °C ± 5 °C	± 3.0
Condition (5)	2.5	2.5	15	30	50	0.5 lag	50 °C ± 5 °C	± 2.0
Condition (6)	5	10	50	100	150	0.5 lag	50 °C ± 5 °C	± 2.0
Condition (7)	0.25	0.25	1.5	3	5	1.0	−20 °C ± 5 °C	± 3.0
Condition (8)	2.5	2.5	15	30	50	1.0	−20 °C ± 5 °C	± 2.0
Condition (9)	5	10	50	100	150	1.0	−20 °C ± 5 °C	± 2.0
Condition (10)	0.5	0.5	3	6	10	0.5 lag	−20 °C ± 5 °C	± 4.0
Condition (11)	2.5	2.5	15	30	50	0.5 lag	−20 °C ± 5 °C	± 3.0
Condition (12)	5	10	50	100	150	0.5 lag	−20 °C ± 5 °C	± 3.0

*When the actual temperature difference between two tests differs from the nominal temperature difference specified for the two tests, the deviation shall be adjusted proportionately.

5.1.8.11 Test No 9: Effect of Variation of Ambient Temperature. The test shall be made at nameplate voltage and rated frequency, and shall be applied to three meters.

The meters shall be placed in a space having a temperature of 23 °C ±5 °C and allowed to stand for not less than 2 hours with the voltage circuits of the meters energized. Reference performance at each of the loads specified in Table 5.1.8.11 shall be obtained after operating the meters for 1 hour at the specified load. The meters shall then be operated and tested at each of the following conditions:

Conditions (1) through (6). These tests shall be made with the meter placed in a space having a temperature of 50 °C ±5 °C. After energizing the voltage circuits of the meters for a period of not less than 2 hours, the appropriate test currents at the power factors listed for conditions (1) through (6) of Table 5.1.8.11 shall be sequentially applied to the meters. Each condition shall be maintained for a period of at least 1 hour before tests are made to determine the deviation from reference performance.

Conditions (7) through (12). Repeat conditions (1) through (6), respectively, except that meters shall be placed in a space having a temperature of –20 °C ± 5 °C.

The effect of variation of temperature upon the performance of the meters shall not exceed that specified in Table 5.1.8.11.

Table 5.1.8.12.1
Effect of Temporary Overloads on Accuracy

Condition	Meter Class			Maximum Deviation in Percent from Reference Performance
	100	200	320	
	(Current in Amperes)			
Reference performance for condition (1)	15	30	50	—
Reference performance for condition (2)	1.5	3	5	—
Condition (1)	15	30	50	± 1.5
Condition (2)	1.5	3	5	± 1.5

5.1.8.12 Test No 10: Effect of Temporary Overloads[11].

5.1.8.12.1 Effect on Accuracy. Self-contained meters shall be subjected to a short-circuit current of 7000 amperes peak, 60 Hz, for not more than 6 cycles (0.1 second). For this test, the current circuits of the meter shall be connected in series adding. The test should be applied to three meters.

The effect of the short-circuit current on the performance of a meter should not exceed that specified in Table 5.1.8.12.1.

In order to eliminate residual effects, it is essential that tests of condition (1) be conducted before tests of condition (2).

5.1.8.12.2 Effect on Magnetic Bearing. A minimum of 0.003 inch down-travel shall remain after the application of the short-circuit current stipulated in 5.1.8.12.1.

5.1.8.12.3 Effect on Mechanical Structure and Insulation. The meter shall withstand, for a duration of 4 cycles, a 60 Hz symmetrical fault current as follows without damage to the mechanical structure or reduction in the insulation level:

Class 100: 10 000 amperes rms
Class 200: 12 000 amperes rms
Class 320: 12 000 amperes rms

5.1.8.13 Test No 11: Internal Meter Losses. The loss in each current circuit of a meter, when tested at rated frequency, at the nameplate test current, and at an ambient temperature of 23 °C ±5 °C, shall not exceed 0.5 voltampere for Class 10 and Class 20 meters or 1.0 volt-ampere for Class 100, Class 200, and Class 320 meters. For two-stator three-phase four-wire-wye Class 10 and Class 20 meters, loss in the current circuit common to both stators shall not exceed 1.0 volt-ampere.

---

[11]This test may be omitted where certified data are furnished by the manufacturer.

Table 5.1.8.14
Temperature-Rise Test Specifications

Meter Class	Wire Size* (AWG Copper)	Current in Amperes	Detachable Meters	
			Socket Rating Amperes	Simulated Meter
10	No 10	10	20 (min)	None
20	No 10	20	20 (min)	None
100	No 2	100	100	Fig 1
200	No 4/0	200	200	Fig 2
320	1 –500 mcm or 2 – 4/0	320	320	Fig 3

*Wire sizes for 100, 200, and 320 amperes are those specified in ANSI/NFPA 70-1981 [7] for 60 °C temperature rating.

The loss in each voltage circuit of a meter, not including auxiliary loads such as voltage indicators, shall not exceed 2.0 watts at rated voltage and frequency at an ambient temperature of 23 °C ±5 °C.

5.1.8.14 Test No 12: Temperature Rise[12]. The test shall be made at nameplate voltage, rated frequency, with the specified current applied to all current circuits in series adding, and at an ambient temperature of 23 °C ±5 °C. This test shall be applied to one meter of each class.

The temperature rise of any of the current-carrying parts of the watthour meter, tested under the specified conditions, shall not exceed 55 °C, except that a higher temperature rise is permissible when suitable insulating materials are used in conformance with the general principles of temperature ratings as specified in IEEE Std 1-1969 [8].

All tests shall be performed with the meter cover in place and in a room essentially free from drafts. The meter shall be mounted in a conventional manner on a suitably rated meter mounting. Not less than 4 feet (8-foot jumper between terminals) of stranded, insulated, copper conductor shall be connected to the line and load-current terminals of the meter or socket. For detachable (Type S) meters the opening where the conductors enter and leave the socket and any other openings shall be closed with suitable material to prevent drafts. The conductor size, test current, and, where applicable, the socket rating and simulated meter are specified in Table 5.1.8.14.

5.1.8.14.1 Test on Class 10 and Class 20 Meters. The temperature-rise test shall be made by determining the increase in resistance of the current circuits.

_____

[12]This test may be omitted where certified data are furnished by the manufacturer.

Before the meter is energized, the resistances of the meter current circuits and the ambient (room) temperature shall be determined. The resistance measurements shall be made by a means capable of determining the change in resistance to an accuracy of ±0.5% or better.

The meter shall be energized at the specified conditions for a minimum period of 2 hours. At the end of the prescribed period, the meter shall be de-energized, noting the time that this action was taken. Resistance readings shall be taken on each current circuit and recorded along with the time at which each measurement was taken. The resistance and time readings shall be repeated until three sets of data have been obtained for each current circuit. These readings shall be taken as quickly as practicable, but in no case should the overall time between de-energization and the last resistance reading exceed 5 min. The temperature rise of the current circuit corresponding to each resistance reading should be calculated by the following formulas:

$$T = 258 \left(\frac{R}{r} - 1 \right) \text{ for copper windings}$$

$$T = 251 \left(\frac{R}{r} - 1 \right) \text{ for aluminum windings}$$

where
T = temperature rise in degrees Celsius
R = hot resistance
r = cold resistance

To determine the temperature rise at the time of de-energization, the temperature rise corresponding to the resistance values for each current circuit shall be plotted against time, and the graph extrapolated to the time of de-energization.

5.1.8.14.2 Test on Class 100, Class 200, and Class 320 Meters. The temperature-rise test shall be made by means of temperature detectors in intimate contact with the metal of the current circuit and located at the approximate electrical center of the current coils.

In the case of meters provided with terminal compartments (Type A), the test shall be conducted under the test conditions specified in 5.1.8.14 until the current-circuit temperatures have stabilized. The temperature rise shall be considered the difference in degrees Celsius between the stabilized temperature and ambient (room) temperature.

For detachable meters (Type S), the test installation shall be standardized using a simulated meter as specified in Table 17 (Figs 1, 2, and 3). The simulated meter shall have the same cover and number of current jumper bars as current circuits in the meter to be tested. A temperature rise shall be determined on the simulated meter by applying the test current to all jumper bars in series until the temperature as indicated by the temperature detector has stabilized. This temperature shall then be recorded and the simulated meter replaced by the meter to be tested. The temperature rise test shall be conducted on the watthour meter under the test conditions specified

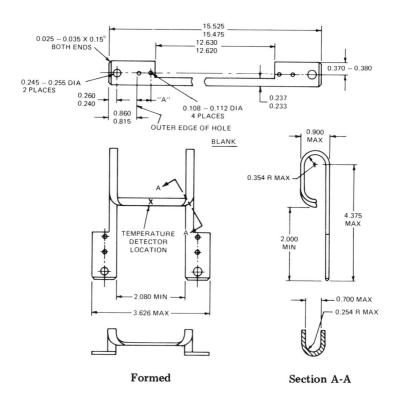

Formed

Section A-A

NOTES: (1) Material is 0.094±0.002 by 0.750±0.005 inch round edge copper with electro tin plate 0.0002—0.0005 inch thick.

(2) Select dimension "A" and retaining lugs or cotter pin holes to suit meter baseplate used.

(3) The temperature detectors shall be so attached, and shall be of such type, that their presence will not appreciably affect the temperature rise of the jumper bars.

(4) All dimensions are in inches.

(5) *Metric Conversion:* Multiply inches by 25.4 to obtain millimeters. Round off to nearest 0.02 millimeter.

Fig 1
Dimensions for Jumper Bars of Simulated
Meter Temperature-Rise Test for Singlephase
and Polyphase Meters.
Maximum Rating 100 Ampere

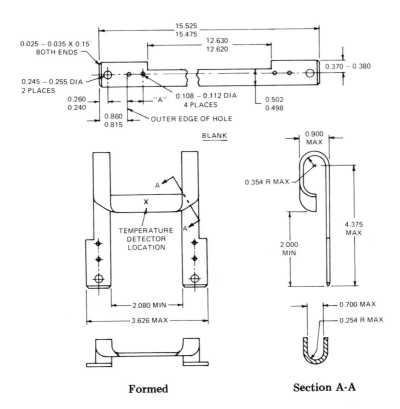

BLANK

Formed

Section A-A

NOTES: (1) Material is 0.094±0.002 by 0.750±0.005 inch round edge copper with electro tin plate 0.0002—0.0005 inch thick.

(2) Select dimension "A" and retaining lugs or cotter pin holes to suit meter baseplate used.

(3) The temperature detectors shall be so attached, and shall be of such type, that their presence will not appreciably affect the temperature rise of the jumper bars.

(4) All dimensions are in inches.

(5) *Metric Conversion:* Multiply inches by 25.4 to obtain millimeters. Round off to nearest 0.02 millimeter.

Fig 2
Dimensions for Jumper Bars of Simulated
Meter Temperature-Rise Test for Singlephase
and Polyphase Meters.
101-to-200-Ampere Rating

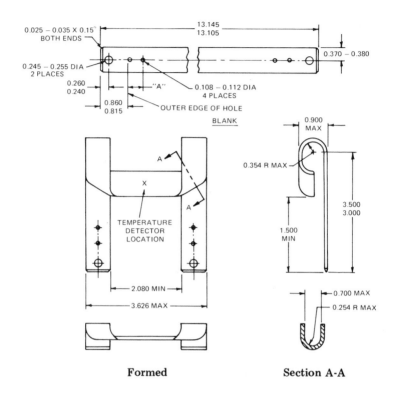

Formed   **Section A-A**

NOTES: (1) Material is 0.094±0.002 by 0.750±0.005 inch round edge copper with electro tin plate 0.0002—0.0005 inch thick.

(2) Select dimension "A" and retaining lugs or cotter pin holes to suit meter baseplate used.

(3) The temperature detectors shall be so attached, and shall be of such type, that their presence will not appreciably affect the temperature rise of the jumper bars.

(4) All dimensions are in inches.

(5) *Metric Conversion:* Multiply inches by 25.4 to obtain millimeters. Round off to nearest 0.02 millimeter.

Fig 3
Dimensions for Jumper Bars of Simulated
Meter Temperature-Rise Test for Singlephase
and Polyphase Meters.
201-to-320-Ampere Rating

in 5.1.8.14. When the temperatures of the meter current circuits have stabilized, the temperatures shall be measured and the empirical temperature rise values of the meter current circuits shall be calculated as follows:

Empirical temperature rise = $\Theta_m - (\Theta_s - 55\ ^\circ C)$

where

Θ_m = measured final temperature rise of current circuit of meter under test

Θ_s = measured final temperature rise of simulated meter jumper bar for the same current phase

NOTE: The foregoing tests shall not be considered conclusive if Θ_s exceeds 65 °C.

5.1.8.15 Test No 13: Effect of Register Friction. The test shall be made at nameplate voltage, rated frequency, 1.0 power factor, and at an ambient temperature of 23 °C ±5 °C.

The change in meter registration after removal of a pointer-type watthour register shall not exceed ±0.5% at 10% of Test Amperes (TA).

5.1.8.16 Test No 14: Effect of Internal Heating. The test shall be made at nameplate voltage, rated frequency, 1.0 power factor, at an ambient temperature of 23 °C ±5 °C with the meter mounted in a conventional manner on a suitably rated meter mounting, and wired with not less than 4 feet of conductor (8-ft jumper between terminals) of a size adequate for the load range of the meter. Openings around the conductor, and any other openings, shall be closed with suitable material to prevent drafts.

The effect of internal heating upon the performance of a meter shall not exceed that specified in Table 5.1.8.16.

Table 5.1.8.16
Effect of Internal Heating

Condition	Meter Class (Current in Amperes)					Maximum Deviation in Percent from Reference Performance
	10	20	100	200	320	
Reference performance for conditions (1), (2), and (7)	10	20	100	200	320	—
Reference performance for conditions (3) and (5)	0.25	0.25	1.5	3	5	—
Reference performance for conditions (4) and (6)	2.5	2.5	15	30	50	—
Condition (1) One-half hour after application of load	10	20	100	200	320	± 1.0
Condition (2) One hour after application of load	10	20	100	200	320	± 1.5

(Continued on Page 87)

Table 5.1.8.16 (*Continued*)
Effect of Internal Heating

Condition	Meter Class (Current in Amperes)					Maximum Deviation in Percent from Reference Performance
	10	20	100	200	320	
Condition (3) Immediately following test for condition (2)	0.25	0.25	1.5	3	5	± 1.5
Condition (4) Immediately following test for condition (3)	2.5	2.5	15	30	50	± 1.5
Condition (5) Two hours after test for condition (4) with meter at no load current during the 2-hour interval	0.25	0.25	1.5	3	5	± 1.5
Condition (6) Immediately following test for condition (5)	2.5	2.5	15	30	50	± 1.0
Condition (7) Immediately following test for condition (6)	10	20	100	200	320	± 1.0

Table 5.1.8.17
Effect of Tilt

Condition	Meter Class (Current in Amperes)					Maximum Deviation in Percent from Reference Performance
	10	20	100	200	320	
Reference performance for conditions (1), (2), (3), and (4)	0.25	0.25	1.5	3	5	—
Condition (1) Top of meter tilted 4 degrees forward	0.25	0.25	1.5	3	5	± 1.0
Condition (2) Top of meter tilted 4 degrees backward	0.25	0.25	1.5	3	5	± 1.0
Condition (3) Top of meter tilted 4 degrees left	0.25	0.25	1.5	3	5	± 1.0
Condition (4) Top of meter tilted 4 degrees right	0.25	0.25	1.5	3	5	± 1.0

(Continued on Page 89)

Table 5.1.8.17 (*Continued*)
Effect of Tilt

| Condition | Meter Class | | | | | Maximum Deviation in Percent from Reference Performance |
| | 10 | 20 | 100 | 200 | 320 | |
			(Current in Amperes)			
Reference performance for conditions (5), (6), (7), and (8)	2.5	2.5	15	30	50	
Condition (5) Top of meter tilted 4 degrees forward	2.5	2.5	15	30	50	± 0.5
Condition (6) Top of meter tilted 4 degrees backward	2.5	2.5	15	30	50	± 0.5
Condition (7) Top of meter tilted 4 degrees left	2.5	2.5	15	30	50	± 0.5
Condition (8) Top of meter tilted 4 degrees right	2.5	2.5	15	30	50	± 0.5

5.1.8.17 Test No 15: Effect of Tilt. The test shall be made at nameplate voltage, rated frequency, 1.0 power factor, and at an ambient temperature of 23 °C ±5 °C, with the top of the meter tilted at an angle of 4 degrees from the vertical: (1) forward, (2) backward, (3) left, and (4) right.

The effect of tilt upon the registration of a meter shall not exceed that specified in Table 5.1.8.17.

Table 5.1.8.18
Effect of Current Surge in Ground Conductor

| Condition | Meter Class | | | Maximum Deviation in Percent from Reference Performance |
| | 100 | 200 | 320 | |
	(Current in Amperes)			
Reference performance	15	30	50	—
Condition (1)	15	30	50	± 1.0

5.1.8.18 Test No 16: Effect of Current Surge in Ground Conductor.[13] Three meters shall be subjected to one transient surge of 20 000 amperes (20 x 50 microsecond wave) of either polarity through a conductor placed vertically 1.5 inches behind the flat portion of the base of the meter, with a socket in place. This test shall not apply for meters with terminal compartments (Type A) or for Class 10 or Class 20 meters. The effect of current surge upon the performance of a meter shall not exceed that specified in Table 5.1.8.18.

5.1.8.19 Test No 17: Stability of Performance[13]. The meters shall be operated continuously at nameplate voltage, rated frequency, and at 10% of test amperes (TA) at 1.0 power factor. The percentage registration shall be determined at an ambient temperature of 23 °C ±5 °C at the start of the test and at 10 successive intervals at least 24 hours apart within a period of 2 weeks.

The change in percentage registration from performance at the start of the test shall not exceed 1.0% on any subsequent test.

5.1.8.20 Test No 18: Independence of Stators

5.1.8.20.1 Two-Stator Meters. The test shall be made on a two-phase circuit at nameplate voltage, rated frequency, and 1.0 power factor.

Throughout the test, the voltage and current circuits of one stator (Stator A) of the meter shall be connected to phase 1 of a two-phase circuit.

For test conditions (1) through (6) the current circuit of the other stator (Stator B) shall not be connected. The voltage circuit of Stator B shall be connected as follows:

Reference performance	Phase 1 direct
Conditions (1) and (2)	Phase 1 reversed
Conditions (3) and (4)	Phase 2 direct
Conditions (5) and (6)	Phase 2 reversed

For test conditions (7) through (12), a current shall be applied to Stator B. The currents in Stator A and Stator B shall be equal in

---

[13]This test may be omitted where certified data are furnished by the manufacturer.

magnitude and each shall be substantially in phase with the voltage
applied to the respective stator. For these test conditions, both the
voltage circuit and the current circuit of Stator B shall be connected
as follows:

Reference performance	Phase 1 direct
Conditions (7) and (8)	Phase 1 reversed
Conditions (9) and (10)	Phase 2 direct
Conditions (11) and (12)	Phase 2 reversed

For two stator three-phase four-wire-wye meters, the current circuit common to both stators shall not be connected. The currents
used shall be twice the values indicated in Table 21. The circuits of
any three-wire stator shall be connected in series and be tested as
one circuit.

The performance of the meter under the test conditions specified
in Table 5.1.8.20.1 shall not deviate from the reference performance
at the corresponding load by an amount exceeding that specified.

5.1.8.20.2 Three-Stator Meters. The test shall be made on a
three-phase four-wire-wye circuit at nameplate voltage, rated frequency, and 1.0 power factor.

Throughout the test, the voltage and current circuits of one stator
(Stator A) of the meter shall be connected to phase 1 of the three-
phase circuit.

For test conditions (1) through (4), the current circuits of the
other stators (Stator B and Stator C) shall not be connected. The
voltage circuit of Stator B and Stator C shall be connected as follows:

Reference performance	— Both Stator B and Stator C, Phase 1 direct
Conditions (1) and (2)	— Stator B, Phase 2 direct; Stator C, Phase 3 direct
Conditions (3) and (4)	— Stator B, Phase 3 direct; Stator C, Phase 2 direct

For test conditions (5) through (8), current shall be applied to the
current circuits of Stator B and Stator C. These currents shall be
equal in magnitude with the current applied to Stator A, and each
shall be substantially in phase with the voltage applied to the respective stator. For these test conditions, both the voltage and current
circuits of Stator B and, similarly, the voltage and current circuits of
Stator C shall be connected as follows:

Reference performance	— Both Stator B and Stator C, Phase 1 direct
Conditions (5) and (6)	— Stator B, Phase 2 direct; Stator C, Phase 3 direct
Conditions (7) and (8)	— Stator B, Phase 3 direct; Stator C, Phase 2 direct

Table 5.1.8.20.1
Test for Independence of Stators in Two-Stator Meters

Condition	Meter Class				Maximum Deviation in Percent from Reference Performance
	10	20	100	200	
	(Current in Amperes)				
Reference performance for conditions (1), (3), and (5)	1	1	6	12	—
for conditions (2), (4), and (6)	5	5	30	60	—
Condition (1)	1	1	6	12	± 1.0
Condition (2)	5	5	30	60	± 1.0
Condition (3)	1	1	6	12	± 1.0
Condition (4)	5	5	30	60	± 1.0
Condition (5)	1	1	6	12	± 1.0
Condition (6)	5	5	30	60	± 1.0
Reference performance for conditions (7), (9), and (11)	0.5	0.5	3	6	—
for conditions (8), (10), and (12)	2.5	2.5	15	30	—
Condition (7)	0.5	0.5	3	6	± 1.0
Condition (8)	2.5	2.5	15	30	± 1.0
Condition (9)	0.5	0.5	3	6	± 1.0
Condition (10)	2.5	2.5	15	30	± 1.0
Condition (11)	0.5	0.5	3	6	± 1.0
Condition (12)	2.5	2.5	15	30	± 1.0

NOTE: For Class 320 meters, the deviation limits given in Table 5.1.8.20.1 are recommended for the following test currents:
Conditions (1) = 20 A, (2) = 100 A, (3) = 20 A, (4) = 100 A, (5) = 20 A, (6) = 100 A, (7) = 10 A, (8) = 50 A, (9) = 10 A, (10) = 50 A, (11) = 10 A, (12) = 50 A.

Table 5.1.8.20.2
Test for Independence of Stators in Three-Stator Meters

Condition	Meter Class				Maximum Deviation in Percent from Reference Performance
	10	20	100	200	
	(Current in Amperes)				
Reference performance for conditions (1) and (3)	1.5	1.5	9	18	—
for conditions (2) and (4)	7.5	7.5	45	90	—
Condition (1)	1.5	1.5	9	18	± 1.0
Condition (2)	7.5	7.5	45	90	± 1.0
Condition (3)	1.5	1.5	9	18	± 1.0
Condition (4)	7.5	7.5	45	90	± 1.0
Reference performance for conditions (5) and (7)	0.5	0.5	3	6	—
for conditions (6) and (8)	2.5	2.5	15	30	—
Condition (5)	0.5	0.5	3	6	± 1.0
Condition (6)	2.5	2.5	15	30	± 1.0
Condition (7)	0.5	0.5	3	6	± 1.0
Condition (8)	2.5	2.5	15	30	± 1.0

NOTE: For Class 320 meters, the deviation limits given in Table 5.1.8.20.2 are recommended for the following test currents:
Conditions (1) = 30 A, (2) = 150 A, (3) = 30 A, (4) = 150 A, (5) = 10 A, (6) = 50 A, (7) = 10 A, (8) = 50 A.

The performance of the meter under test conditions specified in Table 5.1.8.20.2 shall not deviate from the reference performance at the corresponding load by an amount exceeding that specified.

5.2 Demand Meters and Registers
5.2.1 General
5.2.1.1 **Acceptable Demand Meters and Registers.** New types of demand meters and registers, in order to be acceptable, shall conform to certain requirements specified in 5.2.4 through 5.2.11, which are intended to determine their reliability and acceptable accuracy insofar as these qualities can be demonstrated by laboratory tests.

5.2.1.2 **Adequacy of Testing Laboratory.** Tests for determining the acceptability of the types of demand meters and registers under these specifications shall be made in a laboratory having adequate facilities, using instruments of an order of accuracy at least equal to that of the shop instruments and standards described in Section 4. These instruments should be checked against the laboratory secondary standards before and after the tests, or more often as required. The tests shall be conducted only by personnel who have thorough practical and theoretical knowledge of meters and adequate training in the making of precision measurements.

5.2.2 Definitions. For definitions, see Section 2.

5.2.3 Type of Demand Meters and Registers Defined

5.2.3.1 Basic Type. Demand meters or registers are considered to be of the same basic type if they are produced by the same manufacturer, bear a related type designation, are of the same general design, and have the same relationship of parts.

5.2.3.2 Variations within the Basic Type. Demand meters and registers of the same basic type may vary according to the service for which they are designed, such as:

(1) Voltage rating

(2) Class and test-ampere rating

(3) Frequency

(4) Two-, three-, four-, or five-wire

(5) Single-stator or multistator

(6) Wye or delta

(7) Demand interval

(8) Scale capacity

(9) Terminal and mounting arrangement

5.2.3.3 Acceptance of Basic Type in Whole or Part. A basic type of demand meter or register may be accepted as a whole, or a restricted variation of a basic type may be accepted.

5.2.3.4 Minor Variations. Minor variations in the mechanical construction, which are not of such character as to affect the operation of the demand meter or register, may be permitted in different demand meters or registers of the same basic type.

5.2.3.5 Demand Meters and Registers Requiring Separate Tests. Demand meters and registers of the same basic type, but differing within the type as to frequency or number of stators, shall be treated as different types for the purposes of approval tests.

5.2.3.6 Special Types. In the case of a type of demand meter or register that comes within the scope of these specifications but is of such design that some of the tests hereinafter specified are inapplicable or cannot be made under the specified conditions, limited approval may be granted subject to appropriate restrictions.

5.2.3.7 Classification of Demand Meters and Registers. Demand meters and registers presently are divided into two general classifications:

(1) Block-interval demand meters and registers, which record or indicate the demand obtained by arithmetically averaging the watthour meter registration over a regularly repeated time interval

(2) Lagged-demand meters, in which the indication of demand is subject to a characteristic time lag by either mechanical or thermal means

5.2.3.8 Concordance of Demand Meters and Registers. The measurement of demand may be obtained with meters and registers having various operating principles and employing various means of recording or indicating the demand. On a constant load of sufficient duration, accurate demand meters and registers of both classifications will give the same value of maximum demand, within the limits of tolerance specified. On varying loads, the values given by accurate meters and registers of different classifications may differ because of the different underlying principles of the meters themselves. For example, the record of a lagged-demand meter will vary according to the character and sequence of the variations. Furthermore, the records of block-interval demand meters or registers of the same rated time interval may differ if they are not adjusted to reset simultaneously. In commerical practice, the demand of an installation or a system is given with acceptable accuracy by the record or indication of any accurate demand meter or register of acceptable type.

5.2.4 Specifications for Design and Construction

5.2.4.1 Type Designation and Identification. Each demand meter or register shall be designated by type and shall be given a serial number by the manufacturer. The serial number and type designation shall be legibly marked on each demand meter or register. The register ratio shall be marked on the register.

5.2.4.2 Sealing. Demand meters shall be provided with facilities for sealing to prevent unauthorized resetting or entry.

5.2.4.3 Cover. The cover shall be dustproof, and shall be raintight if intended for outdoor installation.

5.2.4.4 Terminals. The terminals of the demand meter shall be so arranged that the possibility of short circuits is minimized when the cover is removed or replaced, connections are made, or the demand meter is adjusted.

5.2.4.5 Construction and Workmanship. Demand meters and registers shall be substantially constructed of good material in a workmanlike manner, with the objective of attaining stability of performance and sustained accuracy over long periods of time and over wide ranges of operating conditions with a minimum of maintenance.

5.2.4.6 Resolution. The resolution of graphic records shall be such that at any point between one-fifth of full scale and full-scale deflection, the demand can be read to within 2% of the full-scale value.

5.2.4.7 Time Characteristics of Lagged Demand Meters. The time required for lagged-demand meters to reach 90% of final indication, with a constant load suddenly applied, shall not be less than 98% of the rated time characteristic.

5.2.5 Selection of Demand Meters and Registers for Approval Tests

5.2.5.1 Samples to Be Representative of Basic Type. The demand meters or registers to be tested shall be representative of the basic type and shall represent the average commercial product of the manufacturer.

5.2.5.2 Number to Be Tested. Eight demand meters or registers shall be subjected to test, except in the case of demand meters or registers of unusual or little-used types, when a smaller number may be taken as being representative of the type, or except as otherwise specified.

When the samples representing a given basic type of demand meter include different current or voltage ratings, there shall be not less than two meters of each of the representative current ratings for each group, all of which preferably should be of the same voltage rating. There shall be not less than two meters of each voltage rating, all of which preferably should be of the same current rating.

When the samples representing a given basic type of demand register include different voltages, demand intervals, or scale capacities, tests shall be made on not less than two registers of each such category.

5.2.5.3 Additional Demand Meters and Registers for Replacements. When practicable, demand meters and registers submitted for approval tests should be accompanied by a sufficient additional number of each variation within the basic type, from which demand meters or registers found defective or those accidentally damaged may be replaced.

5.2.6 Conditions of Test

5.2.6.1 Tests to Be Applied to All Demand Meters and Registers. Each demand meter or register shall be subjected to the tests specified in 5.2.8 through 5.2.11, except that meters or registers that are a modification of a type that has already been subjected to the tests, or demand meters or registers for special services, may be exempted from certain tests.

5.2.6.2 Alternating-Current Tests. All alternating-current tests shall be conducted on a circuit supplied by a sine-wave source with a distortion factor not greater than 3%.

5.2.6.3 Order of Conducting Tests. The items of each test shall be conducted in the order given.

After each change in voltage or load, a sufficient time interval shall be allowed for the demand meter or register to come to a stable condition before making the next observation or test.

5.2.6.4 Specific Conditions of Test. The demand meters or registers shall be mounted on a support free from vibration.

Multistator demand meters shall be tested on a single-phase circuit with the voltage circuits in parallel and all current circuits in series.

All tests shall be made at 23 °C ±5 °C, nameplate voltage, rated frequency, and 1.0 power factor, unless otherwise specified.

5.2.7 Rules Governing the Acceptance of Types

5.2.7.1 Replacements. Replacements or repairs may be made if physical defects of a minor nature become apparent during the tests. If, during the test of a demand meter or register, significant defects in design or manufacture become apparent, the test shall be suspended.

5.2.7.2 Tolerances. Due to possible errors in observations and in the standards employed, a tolerance should be applied to the specified limits of percent deviation for any test condition involving a determination of the accuracy of the demand meter or register. A demand meter or register should not be considered outside of the allowable deviation from the reference performance unless the allowable deviation is exceeded by 0.25%.

5.2.7.3 Basis of Acceptable Performance. A type of demand meter or register shall be acceptable under these specifications when the following requirements are satisfied:

(1) In the tests made on any number of demand meters or registers, not over 30% of such meters and registers shall fail in any one test.

(2) In the tests made on any number of demand meters or registers, the total number of failures shall not exceed 5% of the total number of tests made on all meters or registers.

5.2.8 Performance Requirements Applicable to All Demand Meters and Registers

5.2.8.1 Basis of Performance Determination. For the purpose of approval tests, performance requirements of block-interval demand meters, demand registers, and lagged-demand meters of the mechanical type are based upon the performance of the demand device itself, without regard to the accuracy of the watthour meter with which it is used. The overall accuracy of such a demand meter, then, is the combined accuracy of its register and the electrical element. Performance requirements of the thermal type of lagged-demand meter, however, include the electrical element and the indicating element, since its characteristics are not dependent upon the accuracy of a watthour meter. In either type, however, where a watthour meter is included as part of the demand meter, the watthour meter itself shall meet the requirements of 5.1 of this standard, with suitable allowances for additional losses due to demand circuits.

In the case of demand meters designed and adjusted to be used with specific external instrument transformers or other auxiliary devices, the performance requirements shall apply only to the demand meter proper, with suitable allowances for any special calibration conditions that may have been incorporated in the demand meter. Where the demand meter contains built-in instrument transformers or other special devices forming an integral part

Table 5.2.9.4
Load Performance, Block-Interval Demand Meters
and Registers

Condition	Test Point in Approximate Percent of Full Scale	Permissible Deviation in Percent	
		Demand Interval	Demand
Condition (1)	Any	± 1.0	—
Condition (2)	20	—	± 2.0
Condition (3)	50	—	± 2.0
Condition (4)	90	—	± 2.0

of the demand meter, the performance requirements shall apply to the combination of the demand meter and its built-in auxiliaries.

5.2.9 Performance Requirements — Block-Interval Demand Meters and Registers

5.2.9.1 Initial Conditions. In the tests described in 5.2.9.2 through 5.2.9.7, the tests for demand-interval deviation apply only when a timing element is an integral part of the demand meter or register.

Corrections for any watthour meter errors that may be involved shall be applied when computing demand deviations in the following tests.

Each test shall be made with the demand indicator at zero at the start, and the load of the prescribed amount shall be applied suddenly.

If the meter is of a graphic type, the tests shall be made with the pen or stylus in operating condition and resting with normal pressure against the chart. The proper chart shall be moving at its normal speed, and all results (indications) shall be taken directly from it. Normal speed, when not otherwise specified, shall be the speed corresponding to that indicated on the printed chart or paper.

5.2.9.2 Mechanical Load. The mechanical load imposed on the meter by the demand mechanism shall be within the adjustment range of the meter. This load shall be as nearly constant as practical throughout the entire cycle of operation.

5.2.9.3 Insulation. The insulation between current-carrying parts of separate circuits and between current-carrying parts and other metallic parts shall be capable of withstanding the application of a sinusoidal voltage of 2.5 kV rms, 60 Hz, for 1 minute.

5.2.9.4 Test No 1: Load Performance. Neither the demand-interval deviation nor the demand deviation shall exceed the amounts specified in Table 5.2.9.4.

Table 5.2.9.5
Effect of Variation of Voltage,
Block-Interval Demand Meters and Registers

Condition	Test Point in Approximate Percent of Full Scale	Percent of Nameplate Voltage	Permissible Deviation from Reference in Percent	
			Demand Interval	Demand
Reference performance	90	100	—	—
Condition (1)	90	90	± 1.0	± 1.0
Condition (2)	90	110	± 1.0	± 1.0

5.2.9.5 Test No 2: Effect of Variation of Voltage. Neither the demand-interval deviation nor the demand deviation shall differ from the reference performance by an amount exceeding that specified in Table 5.2.9.5.

Table 5.2.9.6
Effect of Variation of Frequency,
Block-Interval Demand Meters and Registers

Condition	Test Point in Approximate Percent of Full Scale	Percent of Rated Frequency	Permissible Deviation from Reference in Percent Demand Interval
Reference performance	90	100	—
Condition (1)	90	95	± 1.0
Condition (2)	90	105	± 1.0

Table 5.2.9.7
Effect of Variation of Ambient Temperature,
Block-Interval Demand Meters and Registers

Condition	Test Point in Approximate Percent of Full Scale	Ambient Temperature	Permissible Deviation* from Reference in Percent	
			Demand Interval	Demand
Reference performance	50	23 °C ± 5 °C	—	—
Condition (1)	50	50 °C ± 5 °C	± 1.0	± 2.0
Condition (2)	50	−20 °C ± 5 °C	± 1.0	± 2.0

*When the actual temperature difference between two tests differs from the nominal temperature difference specified for the two tests, the change in deviation shall be adjusted proportionately.

5.2.9.6 Test No 3: Effect of Variation of Frequency. The demand-interval deviation shall not differ from the reference performance by an amount exceeding that specified in Table 5.2.9.6.

NOTE: For types of demand meters or registers having a synchronous motor-driven timing mechanism, the foregoing test shall be omitted.

5.2.9.7 Test No 4: Effect of Variation of Ambient Temperature. Three demand meters or registers shall be placed in a space having a temperature of 23 °C ±5 °C, and allowed to stand for not less than 2 hours with the voltage circuits of the meters energized at nameplate voltage. The test load shall then be applied and, after operating for 1 hour, the meters shall be tested. This operation shall be repeated at the various values of temperature shown in Table 5.2.9.7.

5.2.10 Performance Requirements — Lagged-Demand Meters

5.2.10.1 Initial Conditions. Where the demand meter is a component part of the watthour meter, the watthour meter shall be subjected to the tests for acceptance of meters in 5.1 of this standard. The limits of permissible variation for an acceptable meter shall be the same as fixed in 5.1 for a corresponding watthour meter independent of the demand meter, except for loss in watts in current and voltage circuits.

For the tests described in 5.2.10.3 through 5.2.10.8 the demand meters shall be in good operating condition and shall be adjusted as closely as practicable to correct calibration at zero and at some convenient point upscale, preferably at or above 50%. The tests are normally made at nameplate voltage, unless otherwise specified.

Each test shall be made with the demand indicator at zero at the start, and the load of the prescribed amount shall be applied suddenly.

If the meter is of a graphic type, the tests shall be made with the pen or stylus in operating condition and resting with normal pressure against the chart. The proper chart shall be moving at its normal speed, and all results (indications) shall be taken directly from it. Normal speed, when not otherwise specified, shall be the speed corresponding to that indicated on the printed paper chart.

If the meter is of an indicating type, the tests shall be made with the pusher pointer in contact with the maximum demand pointer when up-scale readings are taken, but the pusher pointer shall not be in contact with the maximum demand pointer when on-zero readings of the pusher pointer are taken.

The voltage circuits of thermal demand meters shall be energized for at least 3 hours prior to the initial test and then remain continuously energized for the remainder of the tests.

Tests numbers 2, 3, 4, and 5 (see 5.2.10.4 through 5.2.10.7) do not apply to mechanically lagged meters.

5.2.10.2 Insulation. The insulation between current-carrying parts of separate circuits and between current-carrying parts and other metallic parts shall be capable of withstanding the application of a sinusoidal voltage of 2.5 kV rms, 60 Hz, for 1 minute.

Table 5.2.10.3
Load Performance, Lagged-Demand Meters

Condition	Test Point in Approximate Percent of Full Scale	Permissible Demand Deviation in Percent at End of Four Rated Intervals
Condition (1)	50	± 3.0
Condition (2)	90	± 3.0

NOTE: The time required to reach 90% of final indication is the actual time characteristic of the demand meter and shall not be less than 98% of the rated time characteristic.

Table 5.2.10.4
Effect of Variation of Power Factor,
Lagged-Demand Meters

Condition	Test Point in Approximate Percent of Full Scale	Power Factor	Permissible Deviation from Reference Demand in Percent at End of Four Rated Intervals
Reference performance for condition (1)	50	1.0	—
Reference performance for condition (2)	75	1.0	—
Condition (1)	50	0.5 lag	± 3.0
Condition (2)	75	0.75 lag	± 3.0

5.2.10.3 Test No 1: Load Performance. The demand deviation shall not exceed that specified in Table 5.1.20.3.

5.2.10.4 Test No 2: Effect of Variation of Power Factor. The demand shall not differ from the reference performance by an amount exceeding that specified in Table 5.2.10.4.

Table 5.2.10.5
Effect of Variation of Voltage,
Lagged-Demand Meters

Condition	Test Point in Approximate Percent of Full Scale	Percent of Nameplate Voltage	Permissible Deviation from Reference Demand in Percent at End of Four Rated Intervals
Reference performance for conditions (1) and (3)	50	100	—
Reference performance for conditions (2) and (4)	90	100	—
Condition (1)	50	90	± 1.5
Condition (2)	90	90	± 1.5
Condition (3)	50	110	± 1.5
Condition (4)	90	110	± 1.5

5.2.10.5 Test No 3: Effect of Variation of Voltage. The demand shall not differ from the reference performance by an amount exceeding that specified in Table 5.2.10.5.

Table 5.2.10.6
Effect of Variation of Frequency,
Lagged-Demand Meters

Condition	Test Point in Approximate Percent of Full Scale	Percent Rated Frequency	Permissible Deviation from Reference Demand in Percent at End of Four Rated Intervals
Reference performance for conditions (1) and (3)	50	100	—
Reference performance for conditions (2) and (4)	90	100	—
Condition (1)	50	95	± 1.5
Condition (2)	90	95	± 1.5
Condition (3)	50	105	± 1.5
Condition (4)	90	105	± 1.5

5.2.10.6 Test No 4: Effect of Variation of Frequency. The demand shall not differ from the reference performance by an amount exceeding that specified in Table 5.2.10.6. This test may be omitted except for applications where it is expected that the frequency may vary by more than ±2% from the rated frequency.

Table 5.2.10.7
Equality of Current Circuits, Lagged-Demand Meters

Condition	Test Point in Approximate Percent of Full Scale	Connection of Current Circuits	Permissible Deviation from Reference Demand in Percent at End of Four Rated Intervals
Reference performance	$100/N$*	All circuits in series	—
Condition (1)	$100/N$*	First circuit only	± 2.0
Condition (2)	$100/N$*	Second circuit only	± 2.0
Conditions (3), (4), etc	$100/N$*	Third, fourth, etc, circuit only	± 2.0

*N represents the total number of current circuits in the meter.

5.2.10.7 Test No 5: Equality of Current Circuits. The demand of meters having more than one current circuit, when each current circuit is used separately, shall not differ from the reference performance by an amount exceeding that specified in Table 5.2.10.7.

106

Table 5.2.10.8
Effect of Variation of Ambient Temperature,
Lagged-Demand Meters

Condition	Test Point in Approximate Percent of Full Scale	Power Factor	Ambient Temperature	Permissible Deviation* from Reference Demand in Percent at End of Four Rated Intervals
Reference performance for conditions (1) and (3)	50	1.0	23 °C ± 5 °C	—
Reference performance for conditions (2) and (4)	50	0.5 lag	23 °C ± 5 °C	—
Condition (1)	50	1.0	50 °C ± 5 °C	± 2.0
Condition (2)	50	0.5 lag	50 °C ± 5 °C	± 2.5
Condition (3)	50	1.0	−20 °C ± 5 °C	± 3.5
Condition (4)	50	0.5 lag	−20 °C ± 5 °C	± 4.5

*When the actual temperature difference between two tests differs from the nominal temperature difference specified for the two tests, the change in deviation shall be adjusted proportionately.

5.2.10.8 Test No 6: Effect of Variation of Ambient Temperature. Three demand meters shall be placed in a space having a temperature of 23 °C ±5 °C, and allowed to stand for not less than 3 hours with the voltage circuits of the meters energized at nameplate voltage. The test load shall then be applied for four rated intervals and the indicated demands shall be recorded. This operation shall be repeated at the various values of temperature and power factor given in Table 5.2.10.8. Conditions (2) and (4) do not apply to mechanically lagged demand registers.

Table 5.2.11.1
Test on Timing Element of Chart Drives

Condition	Approximate Percent of Rated Load Applied for 24 Hours	Permissible Timing Deviation in Percent
Condition (1)	25	± 0.25
Condition (2)	100	± 0.25

Table 5.2.11.2
Effect of Variation of Voltage, Chart Drives

Condition	Approximate Percent of Rated Load Applied for 24 Hours	Percent of Nameplate Voltage	Permissible Timing Deviation in Percent
Reference performance	25	100	—
Condition (1)	25	85	± 2.0
Condition (2)	25	110	± 2.0

5.2.11 Tests on Chart Drives for Graphic Demand Meters
    5.2.11.1 Test No 1: Test on Timing Element. Where the timing element is a structural part of the demand meter and provides a continuous record of time on a chart or graph, the chart shall be set to indicate true time, and two records of 24 hours each shall be taken under operating conditions, one at rated load and one at one-quarter rated load. The chart shall be reset to indicate time correctly at the beginning of each 24-hour test. The timing deviation shall not exceed that specified in Table 5.2.11.1.
    5.2.11.2 Test No 2: Effect of Variation of Voltage. The timing deviation shall not differ from the reference performance by an amount exceeding that specified in Table 5.2.11.2.

Table 5.2.11.3
Effect of Variation of Ambient Temperature,
Chart Drives

Condition	Test Point in Approximate Percent of Full Scale	Ambient Temperature	Permissible Timing Deviation* in Percent
Reference performance	50	23 °C ± 5 °C	—
Condition (1)	50	50 °C ± 5 °C	± 2.0
Condition (2)	50	−20 °C ± 5 °C	± 2.0

*When the actual temperature difference between two tests differs from the nominal temperature difference specified for the two tests, the change in the deviation shall be adjusted proportionately.

5.2.11.3 Test No 3: Effect of Variation of Ambient Temperature. Three demand meters shall be placed in a space having a temperature of 23 °C ±5 °C, and allowed to stand for not less than 2 hours with the voltage circuits of the meters energized at approximate nameplate voltage. A record shall then be taken for a 24-hour period. This operation shall be repeated at the values of temperature given in Table 5.2.11.3.

5.3 Pulse Recorders
5.3.1 General
5.3.1.1 Acceptable Pulse Recorders. New types of pulse recorders, in order to be acceptable, shall conform to certain requirements specified in 5.3.8.2 through 5.3.8.5, which are intended to determine their reliability and acceptable accuracy insofar as these qualities can be demonstrated by laboratory tests.

5.3.1.2 Adequacy of Testing Laboratory. Tests for determining the acceptability of the types of pulse recorders under these specifications shall be made in a laboratory having adequate facilities, using instruments of an order of accuracy at least equal to that of the shop instruments and standards described in Section 4. These instruments should be checked against the laboratory secondary standards before and after the tests, or more often as required. The tests shall be conducted only by personnel who have thorough practical and theoretical knowledge and adequate training in the making of precision measurements.

5.3.2 Definitions. For definitions, see Section 2.

5.3.3 Type of Pulse Recorder

5.3.3.1 Basic Type. Pulse recorders are considered to be of the same basic type if they are produced by the same manufacturer, bear a related type designation, are of the same general design, and have the same relationship of parts.

5.3.3.2 Variations Within the Basic Type. Pulse recorders of the same basic type may vary according to the needs of the user, such as (but not limited to) the following:

(1) Voltage
(2) Frequency
(3) Number of channels
(4) Demand interval
(5) Pulse counters
(6) 12- or 24-hour clock
(7) Battery carryover
(8) Type of case
(9) Terminal arrangement

5.3.3.3 Acceptance of Basic Type in Whole or Part. A basic type of pulse recorder may be accepted as a whole, or a restricted variation of a basic type may be accepted.

5.3.3.4 Minor Variations. Minor variations in the mechanical construction, which are not of such character as to affect the operation of the pulse recorder, may be permitted in different pulse recorders of the same basic type.

5.3.4 Specification for Design and Construction

5.3.4.1 Type Designation and Identification. Each pulse recorder shall be designated by type and given a serial number by the manufacturer. The serial number and type designation shall be legibly marked on the frame or chassis of each pulse recorder.

5.3.4.2 Sealing. Pulse recorders shall be provided with facilities for sealing to prevent unauthorized entry.

5.3.4.3 Cover. The cover shall be dustproof, and shall be raintight if intended for outdoor installation.

5.3.4.4 Terminals. The terminals of the pulse recorder shall be so arranged that the possibility of short circuits is minimized when the cover is removed or replaced, connections are made, or tape is replaced.

5.3.4.5 Construction and Workmanship. Pulse recorders shall be substantially constructed of good material in a workmanlike manner, with the objective of attaining stability of performance and sustained accuracy over long periods of time and over wide ranges of operating conditions with a minimum of maintenance.

5.3.5 Selection of Pulse Recorders for Approval Tests

5.3.5.1 Samples to Be Representative of Basic Type. The pulse recorders to be tested shall be representative of the basic type and shall represent the average commercial product of the manufacturer.

5.3.5.2 Number to Be Tested. A minimum of three pulse recorders of a particular type shall be subjected to test.

5.3.5.3 Additional Pulse Recorders for Replacement. When practicable, pulse recorders submitted for tests should be accompanied by a sufficient number of the same type, from which pulse recorders found defective or accidentally damaged may be replaced.

5.3.6 Conditions of Test

5.3.6.1 Tests to Be Applied to All Pulse Recorders. Each pulse recorder shall be subjected to the tests specified in 5.3.8 of this standard except that recorders that are a modification of a type that has already been subjected to the tests may be exempted from certain tests.

5.3.6.2 Order of Conducting Tests. The items of each test shall be conducted in the order given.

5.3.6.3 Specific Conditions of Test. The pulse recorder shall be mounted on a support free from vibration.

All tests shall be made at 23 °C ±5 °C, and at nameplate voltage and frequency, unless otherwise specified.

To avoid ambiguity due to coincidence of data pulses with the timing pulses, it is recommended that the start of the demand interval of the recorders under test be synchronized as closely as practicable, and that the data pulses be started approximately ½ minute after the timing pulse and terminated approximately ½ minute before the end of the demand interval.

Normal commercial alternating-current sources are adequate for test purposes. The pulse source shall be free of transients that will affect pulse count.

5.3.7 Rules Governing the Acceptance of Types

5.3.7.1 Replacements. Replacements or repairs may be made if physical defects of a minor nature become apparent during the tests. If, during the test of a pulse recorder, significant defects in design or manufacture become apparent, the test shall be suspended.

5.3.7.2 Interpretation of Data. Pulse recorders normally require machine reading of their records. In interpreting printout data for the purpose of determining the compliance of records with the performance specifications, care should be taken to avoid misjudging the recorder because of some malfunction of the translating process, computer errors, interchannel crosstalk, defective tape, dirty tape, etc. It is recommended that tests be conducted in such a manner that the maximum practicable redundancy will be obtained. This can be readily achieved by supplying all channels of all recorders from the same pulse source, and monitoring this input with an auxiliary pulse counter. The printed-out data can then be examined with respect to this redundancy for the purpose of avoiding unjustified rejection of the recorder for faults not chargeable to the recorder itself.

Table 5.3.8.3
Load Performance, Pulse Recorders

Condition	Test Point in Approximate Percent of Pulse Capacity	Permissible Deviation in Percent	
		Demand Interval	Pulse Count
Condition (1)	Any	± 0.25	—
Condition (2)	90	—	± 0.25

5.3.7.3 Basis of Acceptable Performance. A type of pulse recorder shall be acceptable under these specifications when the following requirements are satisfied:

(1) In the tests made on three pulse recorders, not more than one recorder shall fail in any one item of any particular test.

(2) No individual pulse recorder shall fail in more than two items.

5.3.8 Performance Requirements

5.3.8.1 Initial Conditions. Tests shall be conducted with the pulse recorder completely assembled in its case and mounted as it would be in normal operation with tape installed. Where magnetic tape is the recording medium, the tape shall be demagnetized before use.

5.3.8.2 Insulation. The insulation between the supply-voltage terminals and the metallic frame or case shall be capable of withstanding the application of a sinusoidal voltage of 1.5 kV rms, 60 Hz, for 1 minute.

NOTE: Low-voltage electronic circuits and the pulse-input (KYZ) terminals are not to be subjected to the insulation test.

5.3.8.3 Test No 1: Load Performance. Neither the demand-interval deviation nor the pulse-count deviation for any demand interval shall not exceed the amounts specified in Table 5.3.8.3. The test shall cover a period of not less than three complete demand intervals for demand-interval deviation, and not less than ten demand intervals for pulse-count deviation. Pulses shall be supplied to all demand data channels simultaneously.

Table 5.3.8.4
Effect of Variation of Voltage, Pulse Recorders

Condition	Test Point in Approximate Percent of Pulse Capacity	Test Point in Percent of Nameplate Voltage	Permissible Change from Reference in Percent	
			Demand Interval	Pulse Count
Reference performance	90	100	—	—
Condition (1)	90	85	± 0.25	± 0.25
Condition (2)	90	110	± 0.25	± 0.25

5.3.8.4 Test No 2: Effect of Variation of Voltage. Neither the changes in demand interval nor in pulse count for any demand interval shall exceed the amounts specified in Table 5.3.8.4. The period covered shall be not less than three complete demand intervals for demand-interval test, and not less than ten demand intervals for pulse-count test. Pulses shall be supplied to all demand data channels simultaneously.

Table 5.3.8.5
Effect of Variation of Ambient Temperature,
Pulse Recorders

Condition	Test Point in Approximate Percent of Pulse Capacity	Ambient Temperature	Permissible Deviation in Percent	
			Demand Interval	Pulse Count
Condition (1)	90	23 °C ± 5 °C	± 0.25	± 0.25
Condition (2)*	90	23 °C ± 5 °C	± 3.0	—
Condition (3)	90	50 °C ± 5 °C	± 0.25	± 0.25
Condition (4)*	90	50 °C ± 5 °C	± 3.0	—
Condition (5)	90	−20 °C ± 5 °C	± 0.25	± 0.25
Condition (6)*	90	−20 °C ± 5 °C	± 3.0	—

*Conditions (2), (4), and (6) are for pulse recorders with battery carryover operating on the dc supply with the ac supply disconnected. The battery shall be placed in the temperature environment during the entire test, but the recorders shall be energized from the ac supply during the conditioning periods. The ac supply is to be disconnected immediately before each demand-interval-deviation test is performed. The battery may be recharged or replaced following each individual temperature run.

5.3.8.5 Test No 3: Effect of Variation of Ambient Temperature. Neither the demand-interval deviation nor the pulse-count deviation for any demand interval shall exceed the amount specified in Table 5.3.8.5. The demand-interval test shall cover a period of not less than three complete demand intervals. The pulse-count-deviation test shall cover not less than ten demand intervals. Pulses shall be supplied to all demand channels simultaneously.

The pulse recorder shall be placed in a space having a temperature of 23 °C ±5 °C, and allowed to stand for not less than 3 hours with nameplate voltage applied to the supply terminals and the specified pulse rate to all channels. After the 3-hour period at this temperature, the test shall be started. This operation shall be repeated at the various values of temperature shown in Table 5.3.8.5.

5.4 Instrument Transformers
5.4.1 General
5.4.1.1 Acceptable Transformers. New types of instrument transformers, in order to be acceptable, shall conform to certain requirements specified in 5.4.4 through 5.4.10, which are intended to determine their reliability and acceptable accuracy insofar as these qualities can be demonstrated by laboratory tests.

5.4.1.2 Adequacy of Testing Laboratory. Tests for determining the acceptability of the types of instrument transformers under these specifications shall be made in a laboratory having adequate facilities, using appropriate laboratory standards, and instruments of an order of accuracy at least equal to that of the shop instruments described in Section 4. The test shall be conducted only by personnel who have thorough practical and theoretical knowledge and adequate training in the making of precision measurements.

5.4.2 Definitions. The following terms are used in this section and are defined in Section 2:

(1) Phase angle of an instrument transformer (see 2.11 and 2.103)

(2) True ratio of an instrument transformer (see 2.41)

(3) Ratio correction factor (see 2.40)

(4) Phase-angle correction factor (see 2.39)

(5) Instrument transformer correction factor (see 2.36)

(6) Continuous thermal current rating factor of a current transformer (see 2.10)

(7) Approval test (see 2.89)

(8) Instrument transformer accuracy class (see 2.33)

5.4.3 Type of Instrument Transformer Defined

5.4.3.1 Basic Type. Instrument transformers are considered to be of the same basic type if they are produced by the same manufacturer, bear a related type designation, are of the same general design, and have the same relationship of parts.

5.4.4 Specifications for Design and Construction

5.4.4.1 Type Designation and Identification. Each transformer shall be designated by type and given a serial number by the manufacturer. The serial number and type designation shall be legibly marked on the nameplate of each transformer.

5.4.4.2 Terminals. The terminals shall be electrically and mechanically suitable for use with aluminum, as well as copper, conductors.

5.4.4.3 Polarity and Terminal Marking. The relative instantaneous polarity of the leads or terminals of instrument transformers shall be clearly indicated by permanent markings that cannot be easily obliterated.

When the polarity is indicated by letters, the letter H shall be used to distinguish the leads or terminals connected to the primary or excited winding; the letter X (also Y and Z if multiple secondary windings are used) shall be used to distinguish the leads or terminals connected to the secondary winding. In addition, each lead shall be numbered, such as: H1, H2, X1, X2. If more than three secondary windings are used they shall be identified X, Y, Z, and W for four windings; X, Y, Z, W, and V for five windings, etc. H1 and X1 (also Y1 and Z1 if used) shall be of the same polarity. When multiple primary windings are provided, the leads or terminals shall be designated by the letter H together with consecutive pairs of numbers (H1, H2, H3, H4, etc). The odd-numbered leads or terminals shall be of the same polarity.

Table 5.4.5.1
Standard Burdens for 5-Ampere
Secondary Current Transformers

Designation of Burden*	Standard Burden Characteristics		Burden Impedance, Voltamperes, and Power Factor for 60 Hz and 5-Ampere Secondary Current		
	Resistance, Ohms	Inductance, Millihenrys	Impedance, Ohms	Volt-Amperes	Power Factor
B-0.1	0.09	0.116	0.1	2.5	0.9
B-0.2	0.18	0.232	0.2	5.0	0.9
B-0.5	0.45	0.580	0.5	12.5	0.9
B-0.9	0.81	1.044	0.9	22.5	0.9
B-1.8	1.62	2.088	1.8	45.0	0.9
B-1.0	0.5	2.3	1.0	25.0	0.5
B-2.0	1.0	4.6	2.0	50.0	0.5
B-4.0	2.0	9.2	4.0	100.0	0.5
B-8.0	4.0	18.4	8.0	200.0	0.5

*Standard burdens for meter applications are B-0.1, B-0.2, B-0.5, B-0.9, and B-1.8.

When taps or leads are provided on the secondary winding or windings, the leads or terminals shall be lettered as required in the preceding paragraph, and numbered X1, X2, X3, etc, or Y1, Y2, Y3, etc, with the lowest and highest numbers indicating the full winding and intermediate numbers indicating the taps in their relative order. When X1 is not used, the lowest number of the two leads used shall be the polarity lead.

5.4.4.4 Construction and Workmanship. Transformers shall be substantially constructed of good material in a workmanlike manner, with the objective of attaining stability of performance and sustained accuracy over long periods of time and over wide ranges of operating conditions with a minimum of maintenance.

5.4.4.5 Nameplates. Transformers shall be provided with nameplates that meet the requirements of ANSI/IEEE C57.13-1978 [4]. In addition, the nameplate shall indicate the standard burdens at which the transformer meets the 0.3 accuracy class.

5.4.5 Burdens and Accuracy Classes of Current Transformers

5.4.5.1 Standard Burdens. Standard burdens for 5-ampere secondary, 60-Hz current transformers shall have resistance and inductance values, together with impedance, voltamperes, and power factors as shown in Table 5.4.5.1 (see ANSI/IEEE C57.13-1978, 6.2, [4]).

Table 5.4.5.2
Standard Accuracy Classes and Corresponding Limits of Transformer Correction Factors for Current Transformers for Metering Service

| Accuracy Class | Limits of Transformer Correction Factor | | | | Limits of Power Factor (Lagging) of Metered Load |
| | 100% Rated Current* | | 10% Rated Current | | |
	Min	Max	Min	Max	
0.3	0.997	1.003	0.994	1.006	0.6 – 1.0
0.6	0.994	1.006	0.988	1.012	0.6 – 1.0
1.2	0.988	1.012	0.976	1.024	0.6 – 1.0

*These limits also apply at the maximum continuous thermal current rating factor.

5.4.5.2 Standard Accuracy Classes. Standard accuracy classes for metering applications are 0.3, 0.6, and 1.2 as shown in Table 5.4.5.2 (see ANSI/IEEE C57.13-1978, 5.3, [4]).

The standard accuracy class for current transformers used in revenue metering is 0.3 accuracy class. If it is not practical to use 0.3-accuracy-class transformers, the transformer correction factors for the transformers to be used should be determined for the secondary burden involved. Corrections should be applied when the correction factors exceed the 0.3-accuracy-class limits. (See 6.1.7.2 and Appendix C.)

Table 5.4.6.1
Standard Burdens for Voltage Transformers

Designation of Burden*	Secondary Voltamperes	Burden Power Factor	Watts	Vars
W	12.5	0.10	1.25	12.4
X	25	0.70	17.5	17.8
M	35	0.20	7.0	34.3
Y	75	0.85	63.8	39.5
Z	200	0.85	170	105
ZZ	400	0.85	340	210

*Standard burdens for meter applications are W, X, M, and Y.

Table 5.4.6.2
Standard Accuracy Classes for Voltage Transformers for Metering Service

Accuracy Class	Limits of Transformer Correction Factor	Limits of Power Factor (Lagging) of Metered Load
0.3	1.003 – 0.997	0.6 – 1.0
0.6	1.006 – 0.994	0.6 – 1.0
1.2	1.012 – 0.988	0.6 – 1.0

5.4.6 Burdens and Accuracy Classes of Voltage Transformers

5.4.6.1 Standard Burdens. Standard burdens for 60-Hz voltage transformers for accuracy rating purposes, expressed in voltamperes, shall be as shown in Table 5.4.6.1. Standard burdens are based on two secondary voltages: 120 and 69.3 volts. (See ANSI/IEEE C57.13-1978 [4], 7.2).

5.4.6.2 Standard Accuracy Class. Standard accuracy classes for metering applications are 0.3, 0.6, and 1.2, as shown in Table 5.4.6.2. (See ANSI/IEEE C57.13-1978, 5.3, [4]).

The standard accuracy class for voltage transformers used in revenue metering is the 0.3 accuracy class. If it is not practical to use 0.3-accuracy-class transformers, the transformer correction factors for the transformers to be used should be determined for the secondary burden involved. Corrections should be applied when the correction factors exceed the 0.3-accuracy-class limits. (See 6.1.7.2 and Appendix C.) Consideration should also be given to corrections for any errors caused by voltage drop in the leads between the voltage transformers and the meters.

5.4.7 Selection of Transformers for Approval Tests

5.4.7.1 Samples to Be Representative of the Basic Type. The transformers to be tested shall be representative of the basic type and shall represent the average commerical product of the manufacturer.

5.4.7.2 Number to Be Tested. Five current or voltage transformers shall be subjected to test, except in the case of very high primary-rated transformers or transformers of unusual or infrequently used types, when a smaller number may be taken as being representative of the type. The samples shall be so selected as to be representative of the range of ratios to be approved.

5.4.8 Conditions of Test

5.4.8.1 Tests to Be Applied to All Transformers. Each transformer shall be subjected to the tests as specified in 5.4.10.

5.4.8.2 Alternating-Current Tests. All alternating-current tests shall be conducted on a circuit supplied by a sine-wave source with a distortion factor not greater than 3%.

5.4.8.3 Order of Conducting Tests. The items of each test shall be conducted in the order given.

5.4.8.4 General. All tests shall be made at a temperature of 23 °C ±5 °C and rated frequency, unless otherwise specified.

In conducting approval tests on instrument transformers, it is recommended that reference be made to ANSI/IEEE C57.13-1978 [4] for detailed test procedures. All tests should be made in accordance with the procedures outlined therein.

5.4.9 Rules Governing the Acceptance of Types

5.4.9.1 Tolerances. The limits of acceptability for current and voltage transformers in each accuracy classification shall include allowances for errors in observation, test equipment variations, and variations in testing techniques. In no case shall a transformer be considered to fall outside the specified limits of an accuracy class unless the limits are exceeded by more than 0.001 in ratio correction factor or 3 minutes in phase angle. Parallelograms that may be used in determining conformance with these accuracy classes are shown in Appendix A of this standard.

5.4.9.2 Basis for Acceptance. A transformer type shall be acceptable under these specifications when the following requirements are satisfied:

(1) Each current and voltage transformer shall withstand all the insulation tests specified in 5.4.10.2.

(2) Each current and voltage transformer shall meet the requirements of both the accuracy and temperature-rise tests specified in 5.4.10.3 through 5.4.10.6.

The failure of a transformer to meet the requirements of the accuracy or temperature-rise tests shall be cause to reject that rating of transformer and other ratings having the same design characteristics. Such rejection shall not preclude the acceptance of the type in other ratings that have different design characteristics and do meet accuracy and temperature-rise requirements.

119

Table 5.4.10.2
Low-Frequency Test Voltages

Insulation** Class, kV	BIL,* kV	Low-Frequency Test,† kV
0.6	10	4
1.2	30	10
2.5	45	15
5.0	60	19
8.7	75	26
15L	95	34
15H	110	34
18	125	40
25	150	50
34.5	200	70
46	250	95
69	350	140
92	450	185
115	550	230
138	650	275
161	750	325
196	900	395
230	1050	460
287	1300	575
400	1800	800
460	2050	920

*Basic impulse insulation level.
**The term "Insulation Class" is no longer in use but is provided here for cross reference to older specifications.

†These test voltages do not apply to insulated-neutral-terminal-type voltage transformers, where the neutral end of the high-voltage winding is insulated for a lower voltage than required for the line terminal. For such transformers, the low-frequency test voltage shall be 19 kV for outdoor types and 10 kV for indoor types. A low-frequency test is not required for grounded-neutral-terminal-type voltage transformers. For information on induced-voltage-test requirements for insulated or grounded-neutral-type voltage transformers, refer to ANSI/IEEE C57.13-1978, 7.9(2), [4].

5.4.10 Performance Requirements
5.4.10.1 Initial Conditions. The transformers shall show no evidence of physical damage, discolored terminals due to overload, or arc tracking on the insulation or bushings.

5.4.10.2 Insulation. The transformers shall not fail when subjected to the following voltage-withstand tests. The tests shall be made either at rated frequency or at a frequency of 60 Hz.

Condition (1). Voltage-withstand test of primary windings to grounded secondary windings and parts, including metallic case, frame, base, nameplate, mounting facilities, and core, if accessible. The test voltage shall be applied for 1 minute between the primary and secondary windings with the secondary windings grounded. The test voltage shall be the low-frequency test voltage as specified in ANSI/IEEE C57.13-1978 [4]. The values are shown in Table 5.4.10.2.

Condition (2). Voltage-withstand test of secondary windings to grounded primary windings and parts as indicated under condition (1). A test voltage of 2.5 kV rms shall be applied for 1 minute between the secondary and primary windings, with all primary windings grounded.

5.4.10.3 Test No 1: Accuracy, Current Transformers. The transformer shall be demagnetized and then tested at 10% and 100% rated primary current and with a primary current equal to the rated current multiplied by the continuous thermal current rating factor. The tests shall be made with the minimum and maximum metering burdens for which a 0.3 accuracy class has been designated by the manufacturer. However, if the intended application of the transformer will involve a burden larger than the maximum metering burden for which the 0.3 accuracy class is specified or if the intended burden will have a power factor nearer to 0.5 than to 0.9, the transformer should also be tested at the standard burden that is most comparable to the intended burden.

In the case of a multiratio transformer, accuracy tests shall be made at ratings representing the highest and the lowest ratios. Sufficient additional tests shall be made to verify the accuracy at all other ratios.

The ratio correction factor and phase angle at each test point in the preceding two paragraphs shall be within the limits of the applicable accuracy class, as specified in 5.4.5.2. (See Fig A1, A2, and A3 of Appendix A.)

5.4.10.4 Test No 2: Temperature Rise, Current Transformers. Each transformer shall be tested with a primary current equal to the rated current multiplied by the continuous thermal current rating factor. The temperature-rise test shall be made in accordance with 6.3.4.2. For an ambient temperature of 30 °C, the winding temperature-rise limits for continuous operation shall not exceed that for which the transformer is rated (55 °C or 80 °C), as determined by the increase in resistance of the winding.

In the case of a multiratio transformer, the temperature-rise test should be performed at the primary rating that is considered to produce the greatest temperature rise.

5.4.10.5 Test No 1: Accuracy, Voltage Transformers. The transformer shall be tested at 90%, 100%, and 110% of rated voltage with zero burden, and at 100% of rated voltage at the maximum burden for which a 0.3 accuracy class is rated by the manufacturer. Accuracy at other burdens may be determined either by test or by means of the circle diagram method. (See Appendix B.)

The accuracy at 90% and 110% of rated voltage at any burden may be determined by applying the ratio and phase-angle differentials determined in the zero burden test to the accuracy for that burden at 100% rated voltage.

The ratio correction factor and phase angle at each test point, and at all burdens for which the transformer is rated 0.3 accuracy class, shall be within the limits of the 0.3 accuracy class, as specified in 5.4.6.2. (See Fig A4 of Appendix A.)

5.4.10.6 Test No 2: Temperature Rise, Voltage Transformers. Each transformer shall be tested with 110% of rated primary voltage applied to the primary winding and with the maximum standard burden for which a 0.3 standard accuracy class is designated by the manufacturer. For an ambient temperature of 30 °C, the temperature-rise limits for continuous operation shall not exceed that for which the transformer is rated (55 °C or 80 °C), as determined by the increase in resistance of the windings.

5.5 Coupling-Capacitor Voltage Transformers

5.5.1 Acceptable Transformer. New types of coupling-capacitor voltage transformers (CCVT) shall, insofar as is practicable, be accepted under the same requirements established for inductive-type voltage transformers specified in 5.4.

6. Test Methods

6.1 Watthour Meters

6.1.1 General

6.1.1.1 Test Requirements. The principal requirements in an acceptable method of testing are accuracy, flexibility, and economy.

6.1.1.2 Accuracy Considerations. The highest practicable accuracy should be obtained in testing. The accuracy of any method of testing is dependent upon a number of factors, which are broadly classified in 6.1.1.2.1 through 6.1.1.2.3.

6.1.1.2.1 Accuracy of Test Standards. The accuracy of an instrument or device used as a standard for testing watthour meters is the accuracy obtainable with reasonable skill under normal conditions of use. The accuracy varies with the type of instrument or device and is affected by various factors, among which are torque, length of scale, frequency variations, waveform, ambient temperature, self-heating, friction, stray fields, inaccuracy of scale markings, and uncertainty of calibration.

6.1.1.2.2 Errors of Observation. Errors of observation are those due to parallax, estimation of fractions of scale divisions, improper evaluation of instrument readings with fluctuating loads, and start and stop errors of standard watthour meters or timing devices.

6.1.1.2.3 Errors in the Method of Test. Errors in the method of testing are those due to improper connections, improper use of standards, improper calculations of measurements, or faulty equipment.

6.1.1.3 Flexibility of Test Equipment. It is important that test equipment be designed to be usable at various current and voltage ratings and for the various types of meters that may be encountered. Future changes in a company's distribution system may require meters of different types and voltages from those in use when equipment is designed. Revisions of national standards for watthour meters may change current ratings or test amperes. Equipment for in-service testing should be designed to be of minimum bulk and weight and should be convenient to use with the types of meter mountings to be encountered.

6.1.1.4 Economic Considerations. Test methods are not recommended that, though highly accurate, may be so expensive or complex that the tests cannot be made with reasonable economy.

6.1.2 Fundamental Methods of Testing

6.1.2.1 Power-Time Methods. In one power-time method, primarily used in the laboratory, the accuracy of the meter under test is determined by measuring with a wattmeter the true power applied to the meter at a substantially constant level over a measured time interval, the total energy through the meter being determined as the product of the constant power and the time interval. Another power-time method in which the power need not be held constant employs a wattmeter and electronic means to obtain the power-time summation.

An essential feature of these methods is that the power-sensing device be available separately for periodic checking as a wattmeter, and that the time-interval device or accumulator also be available separately for checking.

6.1.2.1.1 Advantage. The power-time method of testing has the advantage that it constitutes a direct step in the essential chain of measurements between the local watthour standard and the national electrical standards; that is, the wattmeter calibration can be made potentiometrically in terms of resistance and voltage standards.

6.1.2.1.2 Digital Watthour Meter Comparator. A modification of the power-time method is one in which the power-time integration is mechanically or electrically performed and the result displayed digitally by means incorporated within the device. Such a device may be considered acceptable provided that its overall energy integration is adequate at all required test levels of voltage, current, and power, and that the power-sensing element and the totalizing element are available separately for periodic calibration checks (using normal procedures), and for adjustment where required.

6.1.2.2 Comparative Method. In the comparative method the accuracy of the meter under test is determined by comparison with a reference meter (usually a portable standard watthour meter). It is fundamental to this method that no separate internal checks of the power-sensing element and the accumulator mechanism are possible within the reference meter. If its accuracy is to be established in terms of the national electrical standards, this must be done ultimately in terms of a power-time method.

6.1.2.2.1 Advantage. The comparative method is less susceptible to errors of observation caused by fluctuations in load than is the power-time method. As each test normally requires a number of revolutions of the reference meter, and as these result in the equivalent of a long scale, errors of reading can be of less importance than with the use of an indicating wattmeter.

6.1.2.2.2 Optical-Sensing Test Equipment. Various forms of test equipment are in common use which employ optical-sensing devices to compare the registration of the reference meter with that of the meter under test. Such equipment is used to reduce observation errors, data handling, and time required for testing.

6.1.2.2.3 Automatic Sequenced Testing. Shop-test equipment using the comparative method with optical-sensing devices may be further automated so that heavy load, light load, and 0.5 power factor lag tests may be made automatically in a desired sequence. For large-volume testing, equipment is available that automatically computes and records the percent meter registration.

6.1.3 Methods of Loading. Loading devices suitable for meter testing furnish the means for the tester to operate the meter at the desired test loads. Simplicity of connections, reliability, and minimum weight and volume for in-service testing are determining factors in the selection of the devices. Three available general means of loading are described in 6.1.3.1 through 6.1.3.3.

6.1.3.1 Phantom Loads. Phantom loads consist of low-voltage loading transformers and associated load resistors. These devices, of minimum size and weight, provide a wide range of test currents for watthour meters and also permit testing of high-current-capacity meters on circuits of low primary-ampere rating.

Phantom loads normally provide high-power-factor currents for testing where the exact phase angle is not important. Where such phantom loads are to be used on tests at 0.5 power factor lag, the inherent phase angle of the phantom load should be taken into consideration. Phantom loads, compensated to 1.0 power factor, are available for low-power-factor tests, which eliminate the need for such allowances.

Current at 0.5 power factor lag also may be obtained either by cross-phase connections on a three-phase circuit, or by the addition of reactors to the phantom load. Means of adjusting the final phase angle should be provided where needed.

6.1.3.2 Resistors. Resistors, suitably arranged for convenience in obtaining the desired combinations of loads, have the advantage of simplicity of construction and connection, and in addition provide a true 1.0 power factor load where required. Because of the size and weight, and because the energy is dissipated as heat, use of resistors for in-service testing is normally limited to low-current-capacity meters. For shop testing, resistors are advantageous for special tests and for testing at light loads.

6.1.3.3 Customer Loads. The use of the customer's load is generally to be avoided to prevent misunderstandings and annoyances. It has some application, however, particularly in making tests that have been requested by the customer.

6.1.4 Connections for Shop and In-Service Testing
6.1.4.1 Standard Watthour Meter
6.1.4.1.1 Connections. The voltage circuit of the standard watthour meter shall be connected so that the voltage applied to the standard watthour meter and the meter under test is the same. Voltage-circuit currents shall not pass through current circuits.

The current circuit of the standard watthour meter should be connected in series between the current circuits of the meter under test and the load.

6.1.4.1.2 Selection of Current Circuit. The current circuit of the standard watthour meter should be selected that has a current rating nearest the value of test current to be used. Care should be taken that circuit capacity of the standard watthour meter is not exceeded.

6.1.4.2 Single-Stator Three-Wire Meters. The current circuits are, in general, connected in series and the meter tested as a two-wire meter. The type of test equipment used determines whether or not the meter voltage test link must be opened. In-service testing requires disconnection of all loads except the test load from the meter.

6.1.4.3 Multistator Meters. All tests on multistator meters shall be made with all voltage circuits energized. The four methods described in 6.1.4.3.1 through 6.1.4.3.4 may be used in testing multistator meters. Methods 1 and 2 are recommended as being the more accurate procedures.

6.1.4.3.1 Method 1: Single-Phase Source — Series Test. A balance test may be performed first, if considered necessary, by making separate stator tests with all voltage circuits connected in parallel. Balance tests are made on each stator with test amperes at both 1.0 and 0.5 lagging power factors, and adjustment accomplished with the balance and power factor adjusters.

Following the balance tests, combined stator tests are performed with all current circuits connected in series. Adjustments, when necessary, are accomplished with the full-load, power-factor, and light-load adjusters; however, any change in the individual power-factor adjustments may affect the balance. For two-stator four-wire wye meters, if the current circuit common to both stators is included in the balance tests, it may be omitted from the series test.

6.1.4.3.2 Method 2: Polyphase Source — Balanced Loading Test. This method generally requires the use of as many single-phase standard meters as there are current circuits under test. The meter is connected to a polyphase circuit in the same manner as in service.

A balance test may be performed, if considered necessary, by making single-stator tests. All voltage coils are energized, and the appropriate phase current is applied to each current coil, one at a time. For two-stator four-wire wye meters, the current circuit common to both stators is usually omitted from the test. If the test current and voltage sources allow, tests are made at both unity and 0.5 power factor lag.

After the balance test, all current circuits are energized, the meter is tested, and adjustments, when necessary, are made in the normal manner. For two-stator four-wire wye meters, which have three current circuits and two voltage coils, three single-phase standards are used only if the third phase voltage is available at the test equipment. Alternatively, two standards may be used with an appropriate multiplier if the polyphase currents are closely balanced.

6.1.4.3.3 Method 3: Single-Phase Separate-Stator Test. This method is basically a balance-of-stators test. Because of the elimination of current damping, which would be effective when all circuits are loaded, this test procedure does not produce as accurate overall meter performance as Methods 1 and 2.

For test, the voltage circuits are connected in parallel and each current circuit tested separately. For two-stator four-wire wye meters, the current circuit common to both stators is usually omitted from the test. For two-stator four-wire delta meters, the current circuits of the three-wire stator should be connected in series and treated as a single circuit.

Light-load tests should be made at N times the current value normally used for such tests, where N represents the number of stators in the meter, except for two-stator four-wire wye meters. For the latter, the current circuits that are not common to both stators should be loaded at 4 times the current value normally used, whereas the current circuit which is common to both stators, if tested, should be loaded at 2 times the current value commonly used.

Since light-load adjustment affects all stators when all voltage coils are energized, only one light-load adjustment can be performed. Balance adjusters must be used for independent stator performance adjustment with test amperes at 1.0 power factor, since the full-load adjuster affects the performance of all stators.

6.1.4.3.4 Method 4: Polyphase Separate-Stator Test. The accuracy limitations of Method 3 also apply to this test procedure.

For test, the meter is connected to a polyphase source and each circuit tested separately using the appropriate phase current in each stator. For two-stator four-wire delta meters, the current circuits of the three-wire stator should be connected in series and treated as a single circuit.

Light-load tests should be made at N times the current value normally used for such tests, where N is the number of stators in the meter, except for two-stator four-wire wye meters. For the latter, each circuit should be loaded at 3 times the current value normally used.

The adjustment limitations described under Method 3 also apply to Method 4.

6.1.5 Special Tests for Multicircuit Meters

6.1.5.1 Equality of Current Circuits. Tests should be made on three-wire stators of multistator meters, when any such stators are replaced, to determine the change produced in the performance of the meter by using only one current circuit as compared with that when both current circuits of the stator are used. In such tests, the current applied to the circuits when tested separately should be twice that applied when both current circuits are used.

6.1.5.2 Single-Stator Three-Wire Network Meters. These meters may be tested with either phase-to-phase or phase-to-neutral loading. The first method is preferable, since it always permits the use of one test standard. Meters with 208-volt voltage circuits may be tested on a single-phase supply, whereas meters having 120-volt voltage circuits must be tested on a balanced open-wye supply. To simulate a balanced open-wye supply for testing, an RL circuit may be used with a 208-volt source to obtain 120 volts displaced 30 degrees across the meter voltage coil. With either type of meter, the voltage circuit of the standard watthour meter must be connected phase-to-phase.

6.1.6 Considerations for In-Service Testing

6.1.6.1 Bypass of Meter for Test. The meter should be bypassed when practicable so that the customer will have service during the test.

6.1.6.2 Test Jacks. A test jack may be used for in-service testing of detachable meters provided that it is arranged so that service to the customer will be maintained during test.

6.1.6.3 Disconnection of Load. The customer's load should be effectively disconnected from the meter or test block to preclude the possibility of an error in the results due to a load or fault in the customer's circuit.

6.1.6.4 Demand Devices. When the watthour element of a watthour demand meter is tested, the demand indicator should be moved or set upscale so that the indicator pusher will not cause it to advance during the test. The timing motor on mechanical demand devices should remain energized during the test. After completion of tests, the demand indicator should be moved downscale to its position prior to the test. The indicator pusher of mechanical demand meters should be left at zero.

When an external demand device is used in connection with the watthour meter, precautions should be taken so that the indication of maximum demand will not be influenced by the tests on the watthour meter.

6.1.7 Transformer-Rated Meters

6.1.7.1 Meter Test. The accuracy of a meter installed with current transformers, or with current and voltage transformers, shall be adjusted for use with such transformers except as provided in 6.1.7.2.

6.1.7.2 Separate Tests of Meters and Transformers. The meter may be tested independently of the instrument transformers, provided that the transformer ratios and phase angles have been determined and are taken into account in the adjustment of the meter. The transformer errors may be neglected in the adjustment of the meter if instrument transformers are used that conform to the 0.3-accuracy-class limits at standard burdens approximating the actual secondary burden.

6.1.7.3 Overall Tests. The meter accuracy may be adjusted directly by connecting the test standards in the primary windings of the transformers. In such tests, the secondary burden of the transformers should be the same as in service. Generally the character of the installation is such as to make a primary test difficult or impractical, and therefore it is seldom used.

6.1.7.4 Secondary and Primary Constants. For watthour meters of modern manufacture, the secondary watthour constant (K_h) is shown on the nameplate. The primary watthour constant (PK_h) is the K_h multiplied by the product of the nominal ratios of transformation.

6.1.7.5 Application of Correction Factors. Instrument transformer test cards show the phase angle of the transformer in minutes, positive or negative, and the ratio correction factor (RCF). Information on the application of instrument transformer correction factors is contained in Appendix C.

6.1.8 Average Percentage Registration Determination. The percentage registration of a watthour meter is, in general, different at light loads than at heavy loads, and may have still other values at other loads. The determination of the average percentage registration of a watthour meter is not a simple matter, since it involves the characteristics of the meter and the loading. Various methods are used to determine a single figure that represents the average percentage registration, the method being prescribed by commissions in many cases. Two methods, described in 6.1.8.1 and 6.1.8.2, are used for determining the average percentage registration (commonly called "average accuracy" or "final average accuracy").

6.1.8.1 Method 1. Average percentage registration is the weighted average of the percentage registration at light load (*LL*) and at heavy load (*HL*), giving the heavy load registration a weight of four. By this method:

$$\text{Weighted average percentage registration} = \frac{LL + 4HL}{5}$$

6.1.8.2 Method 2. Average percentage registration is the average of the percentage registration at light load (*LL*) and at heavy load (*HL*). By this method:

$$\text{Average percentage registration} = \frac{LL + HL}{2}$$

6.1.9 Checking Register Ratio and Gear Ratio

6.1.9.1 Checking Register Ratio. The register ratio (R_r) is the number of revolutions of the first gear of the register for one revolution of the first dial pointer. The methods described in 6.1.9.1.1 and 6.1.9.1.2 are applicable for checking register ratios.

6.1.9.1.1 Method 1. Pointer-type or cyclometer-type registers may be compared with a tested register by means of a mechanical device.

6.1.9.1.2 Method 2. For pointer-type registers, the number of revolutions of the first gear of the register required for one revolution of the first dial pointer may be counted. For a register having a large register ratio, the first dial pointer may be advanced only 1/10 of a revolution. When the register ratio contains a fraction, it is desirable that a whole number of revolutions of the first gear of the register be used to give a whole number of revolutions of the first dial pointer. For example, with a register ratio of 27$\frac{7}{9}$, the first gear of the register should be turned 250 times to produce 9 revolutions of the first gear pointer.

6.1.9.2 Checking Gear Ratio. The gear ratio (R_g) is the number of revolutions of the meter rotor for one revolution of the first dial pointer. The following method is applicable for checking gear ratios in the meter shop: After the meters have been tested for accuracy, they may be connected in series with a reference meter and allowed to run for a period of time, such as overnight. The registration of each meter at the end of the period, if the gear ratios are correct, should be the same as that of the reference meter. If voltage circuits cannot be separately energized during the dial run, compensation should be made for errors introduced by the voltage-coil losses.

6.2 Demand Meters, Demand Registers, and Pulse Recorders

6.2.1 General. The test methods described herein are suitable for demand meters, demand registers, and pulse recorders used for measuring kilowatt and kilovar demands, and such other demands as can be properly measured by these devices.

6.2.2 Shop Tests of Block-Interval Pulse-Operated Demand Meters and Pulse Recorders

6.2.2.1 General. When the timing mechanism is an integral part of the demand meter or pulse recorder, it should be tested to determine the correctness of the demand interval. This test may be made at any load point.

Demand meters should be operated at a known number of pulses per demand interval, usually from 30% to 60% of full-scale value. Pulse recorders should be operated at approximately 90% of pulse capacity. It is preferable that this test be run for at least 24 hours.

The pulses may be obtained from a pulse initiator driven by a constant-speed device.

6.2.2.2 Recording Demand Meters. Set the demand indicator at zero and start the tests as outlined in 6.2.2.1 at the beginning of a time interval.

At the conclusion of the tests, the number of pulses recorded should be checked against the number of pulses transmitted per demand interval. If a watthour meter is used to drive the pulse initiator, the sum of the demand readings for all the individual intervals should be checked against the kilowatthour registration during the test. The sum of the demand readings should equal the kilowatthours for a 60-minute demand interval, twice the kilowatthours for a 30-minute demand interval, and four times the kilowatthours for a 15-minute demand interval.

6.2.2.3 Single-Pointer-Form Demand Meters. Allow an interval reset, set the indicating pointer at zero and proceed with the test.

The maximum demand, as determined at the conclusion of the test by reading the indicating pointer to the nearest one-half division, should be checked against the number of pulses transmitted per demand interval. When the demand meter also registers kilowatthours, the testing circuitry should include a counter to register the total number of pulses transmitted during the test in order that the kilowatthours and the kilowatthour value of the total pulses may be compared.

6.2.2.4 Pulse Recorders. Perform necessary tape loading procedures and connections in accordance with manufacturer's instructions and start the tests as outlined in 6.2.2.1 at the beginning of a demand interval. At the conclusion of the tests, the number of pulses recorded should be checked against the number of pulses transmitted per demand interval. If a watthour meter is used to drive the pulse initiator, the sum of the pulses for all the individual intervals should be checked against the kilowatthour registration during the test.

6.2.3 In-Service Tests of Block-Interval Pulse-Operated Demand Meters and Pulse Recorders

6.2.3.1 General. Routine in-service tests are made preferably at the time of tests on the associated watthour meter or meters. The tests should include a check of the electrical and mechanical operation of the demand meter or pulse recorder, an inspection of the pulse initiators, and a check to determine that the demand meter resets properly. The actual demand interval should be determined at any load point.

A demand meter or pulse recorder, its associated pulse initiators, relays, and circuitry may be considered to be operating properly when a kilowatthour check indicates that the demand meter kilowatthours are within acceptable limits of the watthour meter kilowatthours. At least 20 pulses should be transmitted from each pulse initiator during the test and it should be determined that every pulse is received (recorded) by the associated totalizing relay and demand meter. The correctness of the kilowatthour value of a pulse from each pulse initiator should be verified.

6.2.3.2 Recording Demand Meters. Set the demand indicator at zero and start the tests as outlined in 6.2.3.1 at the beginning of a demand interval. At the conclusion of the tests, the number of pulses recorded should be checked against the number of pulses transmitted in a demand interval.

6.2.3.3 Single-Pointer-Form Demand Meters. Allow an interval reset, set the indicating pointer at zero, and proceed with the test. The maximum demand as determined at the conclusion of the test by reading the indicating pointer to the nearest one-half division should be checked against the number of pulses transmitted in a demand interval.

6.2.3.4 Pulse Recorders. Since the pulse count of a pulse recorder may be readily checked against the registration of the corresponding meter at each billing period, in-service tests of pulse recorders are seldom required. When further tests are desired, check the incoming pulses against the counters on the pulse recorder, when available, or against visual or audible test equipment. Where warranted, a test tape may be installed and the reading from the tape compared with the number of incoming pulses.

6.2.4 Shop Tests of Block-Interval Demand Registers

6.2.4.1 General. Shop tests of a pointer-form or a cumulative-form demand register may be made:

(1) With the demand register mounted on a test device, the first gear of the demand register being driven through appropriate gearing by a constant-speed device

(2) With the demand register mounted on a watthour meter

(3) With a manually operated register testing device attached to the demand register.

The third method provides a rapid test on pointer deflection and pusher arm ratio, but does not test the resetting mechanism.

Motor-driven or manually operated test equipment should advance the demand register to a selected test point, usually between 30% and 60% of full-scale value during a demand interval. When the demand register is mounted on a watthour meter for test, the load applied to the watthour meter should be greater than 30% of the demand register full-scale value.

6.2.4.2 Test Methods. The actual demand interval (the time from zero reset to zero reset) should be determined and a billing period reset should be performed to determine that the pointer pusher or test dial pointer returns to zero, after which any of the methods described in 6.2.4.2.1 through 6.2.4.2.3 may be used.

6.2.4.2.1 Method 1. When the register is mounted on a constant-speed test device, the selected scale test point determines the required number of revolutions of the first gear of the register during a demand interval. The maximum demand indication of the register at the end of one or more demand intervals should equal the demand represented by the number of revolutions of the first gear of the register.

6.2.4.2.2 Method 2. When the register is mounted on a watthour meter, the selected scale test point determines the load to be applied to the watthour meter. The number of revolutions of the watthour meter rotor should be determined by an optical sensing device, and the load disconnected when the required number of revolutions has been reached, preferably about a minute before the end of the demand interval. A watthour meter correction for the percentage registration is not necessary when this method is used.

6.2.4.2.3 Method 3. Manually operated register-testing devices are available for quickly checking the kilowatthour gear train and the interval timing gear train.

6.2.5 In-Service Tests of Block-Interval Demand Registers

6.2.5.1 General. When permissible, in-service tests should be made in the shop using any of the methods described in 6.2.4. When tests are required to be made on the customer's premises, the selected scale test point should be above 30% of the demand register scale. When the full-scale value of the demand register exceeds the full-load capability of the associated watthour meter, the full-load capability of the watthour meter should be used in lieu of the full-scale value of the demand register in determining the scale test point.

6.2.5.2 Test Methods. The actual demand interval (the time from zero reset to zero reset) should be determined, and a billing period reset should be performed to determine that the pointer pusher or test dial pointer returns to zero, after which either of the methods described in 6.2.5.2.1 and 6.2.5.2.2 may be used.

6.2.5.2.1 Method 1. A demand register test may be made with the demand register mounted on a watthour meter to which is connected a portable standard watthour meter measuring the same load as that of the watthour meter. The test connections used for watthour meter tests are suitable for demand register tests. When the register is mounted on a multistator meter, the current circuits of the meter should be connected in series when practicable. It is necessary to apply corrections for the percentage registrations of the watthour meter and the portable standard watthour meter when this method is used. It is necessary that the load be applied during a single demand interval, and it may be advantageous to conduct the test for a shorter period of time than that of a full demand interval. The maximum demand indication of the register at the end of the test should be compared with the true demand determined by calculation from readings of the portable standard watthour meter and the demand interval of the demand register.

6.2.5.2.2 Method 2. Manually operated register testing devices are available for quickly checking the kilowatthour gear train and the interval timing gear train.

6.2.6 Shop Tests of Lagged-Demand Meters

6.2.6.1 General. Because of their time-lag characteristics, lagged-demand meters can be tested most economically in the shop by group- or gang-test methods.

6.2.6.2 Zero Test. On thermal-type watt-demand meters, energize only the voltage circuits for eight or more demand intervals and observe the indicating pointer or recording mechanism for proper zero position on the meter scale.

6.2.6.3 Load Test. The two generally accepted methods for load testing of lagged-demand meters are described in 6.2.6.3.1 and 6.2.6.3.2.

6.2.6.3.1 Method 1. Apply accurately and continuously maintained test loads on the demand meter for a period of four or more demand intervals; select appropriate test loads for accuracy test at or near full scale, and for such other scale point tests as may be required. The test loads should be obtained from a stable power source and be maintained at the required values by such manual adjustments as may be indicated by drift on high-accuracy indicating instruments connected in the loading circuit. The test loads may also be maintained continuously at the required values by automatic load-control devices.

6.2.6.3.2 Method 2. Pass the required test loads for a minimum of four demand intervals through the meter under test and through a similar meter that has been previously calibrated to serve as a reference meter. The use of such a reference meter eliminates the necessity for maintaining precisely controlled test load values, as it is only necessary to compare the demand readings of the reference meter and the meter under test.

6.2.7 In-Service Tests of Lagged-Demand Meters

6.2.7.1 General. The shop tests outlined in 6.2.6.2 and 6.2.6.3.1 may also be used for in-service tests. However, since they are time-consuming and may require extensive test equipment, their application for in-service testing is limited.

6.2.7.2 Load Test. Pass the required test load for a minimum of four demand intervals through the meter under test and through a similar meter that has been previously calibrated to serve as a reference meter. The use of such a reference meter eliminates the necessity for maintaining precisely controlled test load values, since it is only necessary to compare the demand readings of the reference meter and the meter under test. As an alternative, the customer's load may be passed through the meter under test and the reference meter for an extended period of time, preferably several days, and thus provide test results based on actual in-service load conditions.

6.3 Instrument Transformers

6.3.1 General. The conventional method for the determination of instrument transformer performance for metering purposes is by means of accurate ratio and phase-angle tests and voltage-withstand tests. The methods described in 6.3.2 through 6.3.4 cover tests, other than factory tests, that are normally made on current and voltage transformers intended to be used for customer billing. It is recommended that reference be made to ANSI/IEEE C57.13-1978 [4], and ANSI/IEEE Std 4-1978 [5] for detailed test procedures.

6.3.2 Insulation

6.3.2.1 General. Other than factory tests, only two voltage-withstand tests normally are made:

(1) Voltage-withstand test of primary windings to grounded secondary windings and parts, including metallic case, frame, base, nameplate, mounting facilities, and core, if accessible

(2) Voltage-withstand test of secondary windings to grounded primary windings and parts as indicated in 6.3.2.1(1)

The test method for each of the foregoing tests is essentially the same, although the test voltage for the first test is determined by the basic impulse insulation level of the transformer, whereas the test voltage for the second test is always 2.5 kV. The tests usually are made at a frequency of 60 Hz. Refer to 6.3.2.4 for percentage of designated test voltages to be used for intial test and for service or periodic test.

Other types of insulation tests that may be made include impulse tests and induced-voltage tests. Refer to ANSI/IEEE C57.13-1978 [4] for details relating to such tests.

6.3.2.2 Test Facilities. Any well-designed high-voltage transformer may be used as a test transformer, but its source of supply should have a capacity of not less than 2 kVA for a voltage of 50 kV or less, and not less than 5 kVA for a voltage greater than 50 kV. Control should be provided so that the test voltage to the instrument transformer under test may be raised and lowered gradually. Full test voltage should not be applied instantaneously to the instrument transformer under test. The test voltage may be measured by any approved method that gives rms values, but preferably by means of a voltage transformer and voltmeter connected to the high-voltage winding of the test transformer. Some protection is desirable in the high-voltage circuit to limit the current in case of failure of the transformer under test.

6.3.2.3 Test Method. All voltage-withstand tests on liquid-immersed transformers should be made with the transformer case properly filled with its insulating liquid.

The terminal ends (and taps, if any) of the winding under test are to be joined together and connected to a line terminal of the high-voltage winding of the loading transformer. The terminals (and taps) of the other windings of the transformer under test and parts as specified in 6.3.2.1(1) are to be grounded and connected to the other terminal of the loading transformer. A convenient arrangement is to place the transformer under test on a grounded metal plate, and connect the grounded plate to terminals (and taps) that are to be grounded. When window-type current transformers are tested, the primary winding should be represented by a conductor in contact with, and substantially conforming to, the inside surface of the window of the transformer (see ANSI/IEEE C57.13-1978, 8.8, [4]).

6.3.4.2 Test Voltage. For the voltage-withstand test of primary windings to grounded secondary windings and parts as specified in 6.3.2.1(1), the test voltage to be used in acceptance tests is shown in Table 5.4.10.1. For voltage-withstand test of secondary windings to grounded primary windings and parts, the test voltage for acceptance test is 2.5 kV. The test voltage to be used in tests of newly purchased transformers should be about 75% of these values. The test voltage to be used in tests of transformers returned from service should be about 65% of these values.

The applied test voltage should be started at one-third or less of full value and increased gradually to full value in not more than 15 seconds. After being held for 1 minute, the voltage should be reduced gradually to one-third of the maximum value or less, in not more than 15 seconds, and the circuit opened.

6.3.3 Accuracy Tests

6.3.3.1 Test Facilities. Either commercial loading devices or special circuits are necessary for comparing primary and secondary values of current or voltage transformers. Standard burdens for current and voltage transformers, and equipment designed to demagnetize current transformers are also necessary.

Additional information on testing instrument transformers is given in ANSI/IEEE C57.13-1978 [4].

6.3.3.2 Current-Transformer Test Methods. Test methods for determining the accuracy of current transformers fall into two general classes, direct and relative. Following demagnetization, accuracy tests of current transformers are usually made by one of the methods described in 6.3.3.2.1 and 6.3.3.2.2, at rated frequency and with an appropriate burden connected to the transformer secondary winding.

6.3.3.2.1 Direct Method. In the direct method, the ratio and phase angle are evaluated in terms of known resistances and reactances. The voltage drops in each of two known noninductive standard resistors that carry the primary and secondary current, respectively, are adjusted so that they are equal and opposite. The ratio of currents equals the reciprocal of the ratio of resistance. The phase angle is determined from the value of standard reactance (inductance or capacitance) necessary to bring the two voltage drops into exact phase opposition.

6.3.3.2.2 Relative Methods. The following two relative methods are commonly used:

(1) *Standard Transformer.* In one relative method the performance of the transformer under test is compared to that of a standard transformer that has been calibrated previously by means of the direct method. This relative method utilizes a bridge circuit in which the constants of the transformer under test are determined by comparison with a calibrated transformer of the same nominal ratio. The secondary windings of the transformers are connected in series. If both transformers have identical characteristics, no current will flow in the detector branch of the bridge circuit. If the characteristics are not exactly equal, the magnitude of the differential current in the detector circuit is used to determine the ratio and phase-angle differences between the transformer under test and the standard transformer. For a current transformer this differential current is measured by adjusting the impedances (resistance and mutual inductance) of a compensating circuit until all of the differential current flows through the compensating circuit instead of through the detector. When this equilibrium is reached, the detector will show zero reading. The arrangement of circuits is such that the scale of the variable resistance may be calibrated directly in ratio values, and the scale of the mutual inductance may be calibrated to read in phase-angle values.

(2) *Current Comparator.* Development of the compensated two-stage current comparator (or its equivalent) has given rise to another basic relative method of increased accuracy capability and ease of operation. The comparator operates as an ampere-turn balance detector, wherein opposing currents in two windings (primary and secondary) acting on a magnetic core are brought to balance by a current in a third winding. (Balance is indicated by a null detector connected to a fourth winding.) When used as a standard to calibrate a current transformer, the primary and secondary windings of the two are connected in series, as in other relative methods, and the third winding of the comparator bridges across the two secondary windings. A network supplies a small injection current through a parallel combination of resistance and capacitance that also bridges the secondary windings. When the RC network is adjusted for a null on the comparator detector, the injection current is equal to the error current of the transformer. Accuracies of a few parts per million are possible.

6.3.3.3 Voltage-Transformer Test Methods. The methods for voltage transformers are essentially the same as those for current transformers. The test set used for the standard-transformer test method is modified so that the phase-angle adjustment is made with a calibrated capacitor instead of the mutual inductor used in the current-transformer test set.

Adaptations of the current comparator are also available for tests of voltage transformers.

6.3.4 Temperature-Rise Tests

6.3.4.1 Test Facilities. A resistance bridge (or bridge network and detector) is required. Calibrated thermometers or thermocouples are required if surface temperature is to be measured.

6.3.4.2 Test Methods. Temperature-rise tests should be made in accordance with ANSI/IEEE C57.13-1978, 8.7, [4]. All temperature-rise tests should be made at rated frequency by determining the increase in the resistance of the windings.

Temperature-rise tests should be made in a room that is essentially free from drafts. Before the transformer is energized, the winding resistance and ambient (room) temperature should be determined. An accurate resistance bridge may be used for resistance measurements. An appropriate burden should be connected to the secondary winding of the transformer. Normally this will be the maximum standard burden for which a 0.3 standard accuracy class is designated by the manufacturer. Since the power factor of the burden used during temperature-rise tests is not important, a resistive burden of the same voltampere rating may be used in place of the maximum rated standard burden. Current transformers may be tested with the secondary winding short-circuited. The temperature-rise test shall continue until constant temperature conditions have been obtained, as determined by surface conditions of the transformer or by noting when the winding resistance becomes constant. The winding resistance measurements should then be made to determine the value at the highest temperature.

Current transformers should be tested by primary loading at a current equal to the ampere rating multiplied by the continuous thermal current rating factor of the transformer. For window-type transformers, the primary conductor used in the test is important. The primary-conductor arrangement used in the test should not constitute an unduly large heat sink or source. It should have a continuous current capacity, in the configuration used, not less than the test current. The capacity should be in accord with the rating established for such conductor arrangement by recognized authority. If more than one primary turn is used, clearance between the transformer body and the encircling primary turns shall be at least 12 inches. For 55 °C-rise transformers, the continuous current capacity of cable used as the primary conductor shall be based on maximum conductor temperature of 75 °C or less. For bus as the primary conductor, the current capacity shall be based on a temperature rise of 50 °C or less. Location of connections to the primary conductor is important from the point of heat transfer and should be a reasonable distance away from the current transformer.

Voltage transformers should be tested with 110% of nameplate voltage applied to the primary winding. Resistance measurements should be made on both the primary and secondary windings. However, measurements on only the secondary winding may be sufficient provided that the transformer remains under test until the temperature becomes stable.

6.3.4.3 Calculation of Temperature. The highest temperature of the windings may be calculated from one of the following formulas:

$$T = \frac{R}{r}(234.5 + t) - 234.5 \text{ (for copper windings)}$$

$$T = \frac{R}{r}(225.0 + t) - 225.0 \text{ (for aluminum windings)}$$

where
    T = temperature in degrees C corresponding to hot resistance R
    t = temperature in degrees C corresponding to cold resistance r
    R = hot resistance
    r = cold resistance

For ambient air temperatures other than 30 °C, the temperature rise $(T - t)$, computed from the appropriate foregoing formula, should be multiplied by:

$$\frac{264.5}{234.5 + \text{ambient air temperature}} \text{ (for copper windings)}$$

or

$$\frac{255.0}{225.0 + \text{ambient air temperature}} \text{ (for aluminum windings)}$$

6.4 Coupling-Capacitor Voltage Transformers

6.4.1 General. The ratio and phase-angle errors of a coupling-capacitor voltage transformer (CCVT) may be determined by utilization of a transportable cascade inductive-type voltage transformer or compressed-gas capacitor as a reference standard or by other acceptable methods.

7. Installation Requirements

7.1 Watthour Meters

7.1.1 Location. The customer or his agent should confer with the supplier of electric service as one of the first steps in planning an electrical installation. The watthour meter should be located where it will be readily accessible and convenient to the supplier of electric service and where it will not be subjected to adverse operating conditions or cause inconvenience to the customer. Normally, the supplier shall determine the location and type of metering equipment to be installed. Watthour meters may be installed in either outdoor or indoor locations. Outdoor installations are generally preferred for convenience to the customer and the supplier.

7.1.2 Rules or Requirements. The supplier of electric service should have available for distribution to customers, architects, contractors, and electricians, copies of rules, specifications, and requirements that may be in force relative to meter installations. Meter installations should conform to the specifications of the supplier as well as to the applicable electrical codes and safety requirements.

7.1.3 Spacing of Meters. When a number of meters are grouped, they should be spaced so that installation, testing, and removal of an individual meter can be accomplished without disturbing adjacent meters.

7.1.4 Identification. When two or more meters are installed in, or on, one building, each meter mounting should be permanently and legibly marked to indicate the customer and service being metered.

7.1.5 Outdoor Installations. Meters and associated metering equipment used on outdoor installations shall be designed specifically for such use, or shall be suitably housed to make the assembly adequate for outdoor service. Meters installed outdoors should not be located where they may be damaged, such as on buildings where unguarded meters will extend into alleys, walkways, or driveways. Meters should be located so they will not be subject to vibration or mechanical damage, and should be mounted without tilt.

Meters installed outdoors should not be more than 6 feet or less than 4 feet above final standing surface, measured from the center of the meter cover. There should be a minimum of 3 feet of unobstructed space in front of the meter, measured from the surface on which the meter is mounted.

Where it is normal practice to remove meters for test rather than test in place, the minimum height may be decreased.

7.1.6 Indoor Installations. Meters installed indoors should be located so that they will not be subject to vibration or mechanical damage, and should be mounted without tilt. On individual installations of meters indoors, the meter should not be more than 6 feet or less than 4 feet above floor level, measured from the center of the meter cover. There should be a minimum of 3 feet of unobstructed space in front of the meter, measured from the surface on which the meter is mounted. On group installations of meters indoors, no meter should be more than 6 feet or less than 3 feet above floor level, measured from the center of the meter cover. There should be a minimum of 3 feet unobstructed space in front of the meter, measured from the surface on which the meter is mounted.

Where it is normal practice to remove meters for test rather than test in place, the minimum height may be decreased.

7.1.7 Wiring for Meters. In general, it is recommended that the meter be installed on the supply side of the service equipment. Any enclosure installed on the supply side of the meter that allows access to the service wiring shall be sealable. When any device or facility of a metering installation is furnished by the customer, such device or facility shall be subject to acceptance by the supplier of electric service.

Where it is necessary to prevent interruption of service to the customer during meter test or changeout, facilities for bypassing the meter, subject to acceptance by the supplier of electric service, shall be provided.

When application is made for service, the customer should definitely understand and agree that the supplier of electric service has the right to inspect all circuits and electric equipment at any reasonable time to ensure that they are properly metered.

7.1.8 Abnormal Service Requirements. Where the electric equipment connected to the supply circuit is of such nature that it will subject the supply circuit or meter installation to abnormal effects, special consideration should be given to the installation of the metering equipment. Whenever the use of such electric equipment is contemplated, its design and installation shall be subject to acceptance by the supplier of electric service.

7.2 Demand Meters, Demand Registers, and Pulse Recorders

7.2.1 Location. Demand meters normally may be installed in locations similar to those required for watthour meters. When equipped with timing mechanisms, inking systems, movements, or tapes that are sensitive to temperature or moisture conditions, the demand meter location should be such that the effect of these conditions will be minimized. Demand meters should be installed at a height that will permit manual reset and maintenance without the use of ladders.

Since some forms of pulse recorders contain sensitive electronic circuits, they are subject to false or lost pulses due to electromagnetic or electrostatic induction. Suitable precautions should be taken to avoid this situation.

7.2.2 Registration. The selection of demand meters and demand devices should be based on the expected maximum demand. The capacity of the demand meter or demand device should be such that the maximum demand registers in the upper half of the scale range, taking into consideration any seasonable characteristics of the customer's load.

7.3 Instrument Transformers

7.3.1 Location. Instrument transformers may be located either indoors or outdoors. Indoor-type transformers may be used outdoors when suitably protected by weatherproof housings. The installation should have adequate electrical and mechanical spacings, and should be properly protected for the type of installation and the voltages involved.

Low-voltage current transformers should be installed as near as practicable to the point of service entrance. It is recommended that current transformers be installed on the line side of the service switch.

7.3.2 Application. The selection of instrument transformers for a specific application should take into consideration the electrical and mechanical designs of the transformers as related to all the characteristics of the circuit on which they are to be applied. All applications should conform to good metering practices. References to short-time mechanical and thermal ratings of current transformers may be found in ANSI/IEEE C57.13-1978 [4], 4.6.

7.3.3 Primary Wiring. It is suggested that not more than one conductor be connected directly to either side of a current-transformer primary. Where multiple conductors are used, they should be connected to a gang terminal or a bus that is connected to the primary of the current transformer. This method of connection provides maximum safety and convenience during installation, replacement, or removal of the current transformer.

7.3.4 Secondary Wiring. Instrument transformer secondary wiring should be in conduit, tubing, harness, or cable separate from all other wiring. The instrument transformer secondary wiring for one service should not be run in the same conduit, tubing, or cable with the wiring of another service. Secondary wiring should be continuous without splices between the transformers and the first meter connection termination, if practicable. It is recommended that color-coded wiring be used. Minimum conductor sizes should be determined on the basis of lengths of circuits, circuit burdens, transformer burden classifications, and metering accuracies required.

7.3.5 Grounding

7.3.5.1 Instrument Transformer Cases. The metal cases, frames, bases, and other mounting facilities of instrument transformers shall be grounded when they are accessible to other than qualified persons, except that the metal cases, frames, bases, and other mounting facilities of current transformers, the primaries of which are not over 150 volts to ground and which are used exclusively to supply current to meters, need not be grounded. (See ANSI/NFPA 70-1981, Section 250-122, [7].)

Nonmetallic surfaces of instrument transformers should be subject to the same safety precautions applicable to their primary circuits.

7.3.5.2 Instrument Transformer Circuits. The secondary circuits of current and voltage instrument transformers shall be grounded when the primary windings are connected to circuits of 300 volts or

141

more to ground and, when on switchboards, shall be grounded irrespective of voltage, except that such circuits need not be grounded when the primary windings are connected to circuits of 750 volts or less and no live parts or wiring are exposed or accessible to other than qualified persons. (See ANSI/NFPA 70-1981, Section 250-121, [7].)

7.3.5.3 Interconnection of Transformer Secondary and Case Grounds. The current- and voltage-transformer secondary wires that are to be grounded should be interconnected and grounded at the same point as the cases, if practicable. The grounded wires of the current- and voltage-transformer secondary windings should be grounded at only one point and as close as practicable to the instrument transformers. The single-point ground on the secondary windings is to prevent accidental paralleling of secondary wires with system grounding wires carrying currents foreign to the metering quantities.

7.3.6 Hazardous Secondary Open-Circuit Voltages. The secondary circuit of a current transformer should not be open when the primary circuit is carrying current because a hazardous voltage may be induced in the secondary circuit under open-circuit conditions. Exception to this safety recommendation may be made only for small-size, through-type current transformers when it has been determined that the peak open-circuit voltage is nonhazardous under the operating or working conditions that will exist.

7.3.7 Paralleling Current-Transformer Secondary Circuits

7.3.7.1 Application. The load output of two or more circuits having a common source may be totalized on one watthour meter by paralleling the secondary circuits of current transformers. The use of this method requires careful consideration of all factors involved to avert excessive metering errors.

7.3.7.2 Effects on Metering Accuracy. Accuracy may be depreciated due to the increase in effective burden on each transformer as a result of paralleling secondary circuits. Accuracy may also be adversely affected if one or more of the paralleled transformers are not carrying load. Accuracy may also be reduced if all load currents to be totalized are not additive.

7.3.7.3 General Requirements. In paralleling the secondary circuits of current transformers, the following requirements and precautions are important from the standpoint of metering accuracy.

(1) All of the transformers shall have the same ratio, and preferably should be the same type.

(2) All transformers that have their secondary windings paralleled shall be connected to the same phase of the primary circuits.

(3) The secondary windings shall be paralleled at the meter to keep the common burden as low as possible.

(4) The secondary windings should be grounded only at the common point at the meter. If it is necessary to ground at the secondary terminals, the conductor for the common return current should be adequately sized.

(5) The meter shall be capable of carrying the combined currents of all the transformers.

(6) The effective burden on each transformer should not exceed its rated burden. Since the total voltage drop across the common burden establishes the flux density for each transformer, the effective burden will increase substantially when a secondary conductor or circuit component carries the secondary current of more than one transformer.

(7) A common voltage for the two or more load circuits shall be available for the meter.

(8) Precaution shall be taken to ensure that neither the secondary nor primary circuits of paralleled current transformers are short-circuited when the metering is in normal operation.

8. Standards for In-Service Performance

8.1 Watthour Meters

8.1.1 General. The purpose of this section is to prescribe limits of accuracy for alternating-current meters being used or to be used for the revenue metering of electric energy and to outline meter test and inspection procedures that will reasonably assure compliance with the requirements of this section. The tests and inspections set forth in Section 8 are in addition to the type acceptance tests outlined in Section 5, Acceptable Performance of New Types of Electricity Meters and Instrument Transformers.

8.1.2 Definitions. The following terms are used in this section and are defined in Section 2, Definitions:

(1) Inspection — meter installation (see 2.31)

(2) Meter shop (see 2.80)

(3) In-service test (see 2.91)

(4) Referee test (see 2.92)

(5) Request test (see 2.93)

(6) Watthour meter — creep (see 2.110)

(7) Watthour meter — test current (TA) (see 2.133)

8.1.3 Accuracy Requirements

8.1.3.1 General. No meter shall be placed in service, or allowed to remain in service, that has an incorrect register constant, watthour constant, gear ratio, or dial train; is mechanically or electrically defective, incorrectly connected, installed, or applied; or registers outside the limits specified in 8.1.3.4.

8.1.3.2 Test Loads. For self-contained meters, heavy load shall be approximately 100% of test current and light load approximately 10% of test current. For meters used with current transformers, heavy load shall be approximately 100% of either meter test current or the secondary current rating of the current transformers; light load shall be approximately 10% of the selected heavy-load current.

8.1.3.3 Acceptable Performance. Although compliance with the provisions of 8.1.3.4 is mandatory, the performance of a watthour meter is considered to be acceptable for 8.1.8 when the meter disk does not creep, and when the percentage registration is not more than 102% or less than 98%, calculated in accordance with one of the methods described in 6.1.8.

Where instrument transformers are used in conjunction with the meter, the registration limits apply to the meter equipment as a whole, except as provided in 6.1.7.2.

8.1.3.4 Adjustment Limits. When a test of a watthour meter indicates that the error in registration exceeds 1% at either light load or heavy load at unity power factor, or exceeds 2% at heavy load at approximately 0.5 power factor lag, the percentage registration of the meter shall be adjusted to within these limits of error, as closely as practicable to the condition of zero error. Where instrument transformers are used in conjunction with the meter, these limits

apply to the meter equipment as a whole, except as provided in
6.1.7.2. All meters that are tested shall be left without creep.

8.1.4 Tests

8.1.4.1 Preinstallation Tests. New meters that are tested prior to
installation shall be tested at heavy load and at light load at unity
power factor, at heavy load at approximately 0.5 power factor lag,
and for creep if required by 8.1.4.4.

8.1.4.2 As-Found Service Tests. Meters that are tested to determine their performance in service shall be tested at heavy load and at
light load at unity power factor. For referee and request tests, a test
for creep should generally be made. In other cases, a creep test shall
be made if required by 8.1.4.4.

8.1.4.3 As-Left Tests. Meters that are to remain in service after
adjustment, or that are tested to determine their performance before
reinstallation, shall be tested at heavy load and at light load at unity
power factor, and for creep if required by 8.1.4.4. Meters on which a
part, or all, of the electromagnetic structure is altered or replaced
shall also be tested at heavy load at approximately 0.5 power factor
lag.

8.1.4.4 Test for Creep. If percent registration at light load deviates by 2% or more from registration at heavy load, a test for creep
shall be made. For the purpose of this test, a meter is considered to
creep if, with the load wires removed and with normal operating
voltage applied to the voltage circuits of the meter, the rotor makes
one complete revolution in 10 minutes or less.

8.1.5 New Meters

8.1.5.1 General. New meters shall be inspected and tested in a
meter shop or laboratory, either on a 100% basis or on a statistical
sampling basis acceptable to the regulatory authority, and appropriate action shall be taken to assure that the meters conform to the
requirements of 8.1.3.

8.1.6 Meters on Customer's Premises

8.1.6.1 Inspection. An inspection of the meter and auxiliary
equipment shall be made before the meter is tested. An inspection
and suitable checks shall also be made following installation when
instrument transformers, phase-shifting transformers, or pulse initiators are a part of a metering installation, to assure that the installation is functioning properly.

8.1.6.2 Tests

8.1.6.2.1 As-Found Tests. An as-found test, performed on the
customer's premises to determine the in-service meter performance,
shall be made in accordance with 8.1.4.2 before removing the meter
cover.

8.1.6.2.2 As-Left Tests. If the meter is to remain in service, an
as-left test shall be made in accordance with 8.1.4.3, and the meter
shall be adjusted, if necessary, to conform to the requirements of
8.1.3.4.

8.1.7 Meters Removed from Service

8.1.7.1 Inspection. An inspection should be made prior to the removal of the meter.

8.1.7.2 Tests

8.1.7.2.1 As-Found Tests. Immediately prior to or following removal of the meter from the customer's premises, as-found tests should be made on the meter in accordance with 8.1.4.2, before removing the meter cover if practicable, and before making any adjustment.

8.1.7.2.2 As-Left Tests. If the meter is to be reinstalled in service, as-left tests should be made in accordance with 8.1.4.3. If tested, the meter shall be adjusted as necessary to conform to the requirements of 8.1.3.4.

8.1.8 In-Service Performance Tests

8.1.8.1 General. Self-contained single-phase meters, self-contained polyphase meters, and three-wire network meters not equipped with demand registers or pulse initiators may be tested under any of the programs listed in 8.1.8.2. In each of these three categories all of the meters shall be tested under the same program, except that when polyphase meters are being tested under a variable-interval or statistical sampling program, small homogeneous groups of such meters may be tested under the periodic test schedule if this requires less testing. Meters equipped with demand registers or pulse initiators shall be tested in accordance with 8.2.3.1. All other meters shall be tested in accordance with 8.1.8.4.

In-service tests may be made on the customer's premises or in the utility's meter shop. However, it is recommended that bottom-connected meters associated with instrument transformers, or phase-shifting transformers, or those having pulse initiators, be tested on the customer's premises.

Tests made for other purposes, such as request or referee tests, shall not be considered as in-service performance tests, except for those meters tested as specified in 8.1.8.4.

8.1.8.2 Test Programs.[14] Permissive test programs referred to in 8.1.8.1 shall be as follows:

(1) Periodic interval (see 8.1.8.4)

(2) Variable interval (see 8.1.8.5)

(3) Statistical sampling (see 8.1.8.6)

8.1.8.3 Objectives. The primary purpose of in-service performance testing is to provide information on which the utility may base a program to maintain meters in an acceptable degree of accuracy throughout their service life. The testing program, maintenance procedures, meter design, and the level of accuracy specified must be such that a realistic balance exists between the benefits realized from high accuracy levels and the cost of achieving these levels. The three alternate programs of in-service performance specified in 8.1.8.2 for

---

[14]For self-contained single-phase meters, self-contained polyphase meters, and three-wire network meters it is expected that the periodic test schedule will be used only by utilities with a very small number of such meters. The variable-interval plan will probably be used by utilities of intermediate size, and by those that do not want to be burdened with the complexities of a statistical sampling plan.

self-contained single-phase meters, self-contained polyphase meters, and three-wire network meters are discussed in 8.1.8.4 through 8.1.8.6.

The periodic test schedule provides for a fixed interval between tests of self-contained single-phase meters, self-contained polyphase meters, and three-wire network meters. Although the periodic test schedule may achieve the objective of maintenance of an acceptable degree of accuracy on the average, under some circumstances it will not meet the objective of achieving these results for particular meter groups or, when it does meet the objectives for a particular meter group, the cost may not be consistent with the value of the benefits received. The chief weaknesses of the periodic test schedule are that it: (1) fails to recognize the differences in accuracy characteristics of various types of meters; and (2) fails to provide incentives for maintenance and modernization programs.

The variable-interval plan provides for the division of meters into homogeneous groups and the establishment of a testing rate for each group based on the results of in-service performance tests made on meters longest in service without test. The maximum test rate recommended is 25% per year. The minimum test rate recommended provides for the testing of a sufficient number of meters to provide adequate data to determine the test rate for the succeeding year. The provisions of the variable-interval plan recognize the differences between various meter types and encourage adequate meter maintenance and replacement programs.

The statistical sampling program included herein has purposely not been limited to a specific method, since it is recognized that there are many acceptable ways of achieving good results. The general provisions of the statistical sampling program provide for the division of meters into homogeneous groups, the annual selection and testing of a random sample of meters of each group, and the evaluation of the test results. The program provides for accelerated testing, maintenance, or replacement if the analysis of the sample test data indicates that a group of meters does not meet the performance criteria.

In determining the results of the maintenance program by statistical sampling methods, the utility assumes certain predetermined risks that a good group of meters will appear to be bad and thereby require increased maintenance unnecessarily. The customer also assumes certain predetermined risks that a bad group of meters will appear to be good and will not be subjected to increased maintenance, as they should. For a stable group, the chance that such an incorrect decision will be repeated in the second and third year diminishes as the square and cube, respectively, of the chance that it occurred the first year. In other words, if the chance is one in ten that an incorrect decision occurred in one year's test, then the chance that it would be repeated on the same group of meters for the next year is only one in a hundred. The chance that it would be repeated the third successive year would be one in a thousand.

8.1.8.4 Periodic Test Schedule. In the test intervals specified in 8.1.8.4(1) and 8.1.8.4(2) the word "years" means calendar years.

The periods stated are recommended test intervals. There may be situations in which individual meters, groups of meters, or types of meters should be tested more frequently. In addition, because of the complexity of installations using instrument transformers and the importance of large loads, more frequent inspection and test of such installations may be desirable.

In general, periodic test schedules should be as follows:

(1) Meters with surge-proof magnets and without demand registers or pulse initiators: 16 years

(2) Meters without surge-proof magnets and without demand registers or pulse initiators: 8 years

NOTE: For test schedules for meters with demand registers or pulse initiators, see 8.2.3.1.

8.1.8.5 Variable-Interval Plan. A variable-interval plan for testing self-contained single-phase meters, self-contained polyphase meters, and three-wire network meters, conforming to the general provisions set forth in 8.1.8.5, may be used if acceptable to the regulatory authority.

The meters shall be divided into homogeneous groups, such as manufacturers' types, and may be further subdivided in accordance with location or other factors that may be disclosed by test records to have an effect on the percentage registration of the meters. Subsequently, groupings may be modified or combined if justified by the performance records.

The meters to be tested shall be representative of those longest in service without test.

The percentage of meters to be tested in each group during the current year shall be dependent upon the results of the in-service performance tests made during the preceding year, or years, up to a maximum of three. The test rate or percentage of meters to be tested in each group shall be a function of the percentage of meters found outside the acceptable performance limits specified in 8.1.3.3. The relationship used to determine the test rate from the test data shall be designed to achieve the objectives set forth in 8.1.8.3 and shall provide for increasing test rates with increases in the percentage of meters outside the acceptable limits. It is recommended that the formulas used should provide for a test rate of 12.5% when the percentage of meters outside the acceptable limits equals 3.0%, and for a maximum test rate of 25.0% when the percentage of meters outside the acceptable limits equals or exceeds 6.0% for a nonlinear relationship, or equals or exceeds 5.0% for a linear relationship. It is also recommended that the minimum number of meters tested in each group should be 200, or 12.5%, whichever is the lesser.

Formulas resulting in either linear or nonlinear relationships that satisfy the general requirements of 8.1.8.5 are acceptable. Such formulas may contain provisions for increasing the test rate as a function of the percentage of meters that register in excess of 102%.

A variable-interval plan shall be accompanied by a liberal policy of making request tests and a procedure whereby unusually high or low bills for service would be detected and investigated.

Records shall be maintained and tabulated to indicate the number of meters in each homogeneous group in service at the beginning of each year, the number of meters tested for each homogeneous group, the test results for each group, and any necessary corrective action taken.

8.1.8.6 Statistical Sampling. A statistical sampling program for self-contained single-phase meters, self-contained polyphase meters, and three-wire network meters, conforming to the general provisions set forth in 8.1.8.6, may be used if acceptable to the regulatory authority. The program used shall conform to accepted principles of statistical sampling based on either variables or attributes methods, and should be evaluated by independent mathematical statisticians.

A statistical sampling program shall include an adequate policy for testing meters on request and a procedure whereby unusually high or low bills for service would be detected and investigated.

The meters shall be divided into homogeneous groups, such as manufacturers' types, and may be further subdivided in accordance with location or other factors that may be disclosed by test records to have an effect on the percentage registration of the meters. Subsequently, groupings may be modified or combined if justified by the performance records.

A sample shall be taken each year from each homogeneous group.

It is extremely important that each meter in the sample be drawn at random; that is, every meter in the group must have an equal chance to be drawn. In order to accomplish this aim it is advisable to use a table of random numbers.

The sample taken each year shall be of sufficient size to demonstrate, with reasonable assurance, the condition of the group from which the sample is drawn.

The sampling program shall be designed to accomplish the objectives set forth in 8.1.8.3, and it should contain a table of mathematically calculated sample sizes and related constants for determining the characteristics of the homogeneous group, accompanied by curves for determining the risk of making an incorrect decision.

If a group of meters does not meet the performance criteria, then corrective action shall be taken. This action may consist of an accelerated test program to raise the accuracy performance of the group to acceptable standards, or it may consist of removing the group from service. An accelerated test program should provide for testing at rates that vary in accordance with the calculated percentage of meters outside the acceptable limits of accuracy in rejected groups. In its application to an individual group, the rate of testing or removal from service should be such that the required corrective action is completed within 4 years. Accelerated testing may be discontinued when the test results indicate that the rejected group is within acceptable limits.

Records shall be maintained and tabulated to indicate the number of meters in each homogeneous group in service at the beginning of each year, the number of meters making up the sample for each homogeneous group, the test results for each group, and any necessary corrective action taken.

8.2 Demand Meters, Demand Registers, and Pulse Recorders

8.2.1 Accuracy Requirements

8.2.1.1 Acceptable Performance. The performance of a demand meter or register shall be acceptable when the error in registration does not exceed 4% in terms of full-scale value when tested at any point between 25% and 100% of full-scale value.

Under usual operating conditions, the performance of a pulse recorder shall be acceptable when the monthly kilowatthours calculated from the pulse count do not differ by more than 2% from the corresponding kilowatthour meter registration and the timing-element error is no more than ±2 minutes per day.

8.2.1.2 Test Points. Demand meters and registers should be tested at load points at or above 50% of full scale. However, they may be tested at another scale point if conditions warrant.

8.2.1.3 Adjustment Limits. When a test of a demand meter or register indicates that the error in registration exceeds ±4% in terms of full-scale value, the demand meter or register shall be adjusted to within ±2% of full-scale value.

When a timing element also serves to keep a record of the time of day at which the demand occurs, it shall be corrected if it is found to be in error by more than ±2 minutes per day.

For pulse recorders having no adjustments, errors in their pulse count usually will be due to malfunction of associated equipment or of one or more components of the recorder. Correction of the offending condition or replacement of defective parts is required.

8.2.1.4 General. Demand meters or registers that are found to register due to any influence not properly caused by the actuating mechanism shall be corrected or removed from service.

Block-interval demand meters or registers that do not properly reset to zero shall be corrected or removed from service.

Pulse initiators that are found to transmit improper pulses or an incorrect number of pulses shall be corrected or removed from service.

Pulse recorders that are found to be defective or to record an incorrect number of pulses shall be corrected or removed from service.

8.2.2 Test Rules

8.2.2.1 Initial Test. All demand meters, registers, and pulse recorders shall be tested prior to installation or within 60 days after installation.

8.2.2.2 Place of Test. Tests may be made on the premises where the demand meter or register is installed, or the demand meter or register may be removed to the meter shop for test.

8.2.2.3 Comparison in Lieu of Test. In testing a block-interval pulse-operated recording demand meter, a comparison of the summation of the demands recorded for any specific period of time with the kilowatthour registration of the associated watthour meter, during the same period of time, may be considered as a test of the demand meter or pulse recorder.

8.2.2.4 Pulse Initiators. When it is possible to check the operation of pulse initiators by means of counters installed for the purpose, no other test need be made of the pulse initiators or their circuits, but the demand meter or register shall be tested as prescribed.

8.2.2.5 Transmission of Pulses. When a demand meter, demand register, or pulse recorder is actuated by pulses from pulse initiators installed in one or more watthour meters or relays, each of the associated watthour meters or relays shall be caused to transmit a minimum of 20 pulses to the demand meter or register or suitable counting device as a check on the pulse initiators, associated equipment, and circuitry. The demand meter or register shall be tested as prescribed.

8.2.2.6 Mechanically Actuated Demand Devices. When a demand meter or register is mechanically actuated by a watthour meter or demand-totalizing relay, the test shall be made using the watthour meter, relay, or proper equivalent driving force to actuate the demand element.

8.2.3 In-Service Performance Tests

8.2.3.1 Periodic Test Schedule. Periodic tests should be made with sufficient frequency to ensure continued reliability and acceptable accuracy of the demand metering system as a whole. The proper periodic test interval will depend upon the inherent reliability of the particular type of demand meter and associated equipment as determined by past experience. In addition, because of the complexity of some installations using demand devices, and the importance of large loads, more frequent inspection or test of such installations may be desirable.

In general, periodic test schedules should be as follows:

(1) Block-interval demand-register-equipped watthour meters:
  (a) Meters with surge-proof magnets: 12 years
  (b) Meters without surge-proof magnets: 8 years
(2) Block-interval graphic watthour demand meters: 2 years
(3) Lagged-demand meters: 8 years
(4) Pulse recorders and pulse-operated demand meters in combination with pulse-initiator-equipped watthour meters: 2 years

If a comparison is made between the meter registration and the recorder registration each billing period, and the recorder registration agrees within 1% of that registered by the associated meter, the schedule for pulse recorders and pulse-operated demand meters should be as follows:

  (a) Meters with surge-proof magnets: 16 years
  (b) Meters without surge-proof magnets: 8 years

If recorder-meter registration checks do not agree within 1%, the demand metering equipment should be tested.[15]

8.3 Instrument Transformers

8.3.1 General. No instrument transformer shall be placed in service, or allowed to remain in service, if it shows evidence of physical damage, discolored terminals due to overload, change in texture or resiliency of insulation, or arc tracking on the insulation or bushings.

8.3.2 Test Rules

8.3.2.1 Preinstallation Tests. Prior to installation, all new instrument transformers shall be tested for voltage withstand and for ratio correction factor and phase angle, in the meter shop or by the manufacturer.

Current transformers should be tested at rated frequency, at 10% and 100% of rated primary current, and with the maximum burden for which a 0.3-accuracy-class rating is given by the manufacturer, or at the intended operating burden or nearest standard burden thereto. Voltage transformers should be tested at rated frequency and at such voltages and burdens as to give assurance of satisfactory in-service performance. Voltage-withstand tests should be made at 75% of the voltages specified in Table 5.4.10.2. A record should be made of all test results.

8.3.2.2 Transformers Removed from Service. An instrument transformer that has been removed from service should be tested for accuracy or voltage withstand, or both, prior to reinstallation if the reason for removal, or physical appearance, or record of performance gives cause to doubt its reliability. The tests should conform with 8.3.2.1, except that the voltage-withstand tests should be made at 65% of the values specified in Table 5.4.10.2. A record should be made of all test results.

8.3.3 In-Service Performance Tests

8.3.3.1 Periodic Test Schedules. Experience has demonstrated that instrument transformers in service maintain their accuracies except when the transformers have been severely overloaded for extended periods of time, have become physically damaged, or have been subjected to abnormal conditions. Consequently, the periodic testing of instrument transformers is considered to be unnecessary.

---

[15]In practice, agreement is sought to within 1 or 1½ units of the least significant dial. This implies that the watthour meter constant (K_r) should be taken into account. Also, when metering transformers are used, calculations should involve the primary kilowatthours as indicated in the following formula:

$$\frac{\left(\begin{array}{c}\text{Primary kWh from}\\ \text{watthour meter}\end{array}\right) - \left(\begin{array}{c}\text{Primary kWh from}\\ \text{total pulse count}\end{array}\right)}{\text{Watthour Meter Register Constant}} \leqslant 1.5$$

8.3.3.2 In-Service Inspection. When metering installations are inspected on periodic schedules, the instrument transformers associated with the installations should receive a close visual inspection for correctness of connections and evidence of any damage as outlined in 8.3.1.

8.3.3.3 Heavy Burden Test. Current transformers may be tested with a suitable variable-burden device to determine whether the windings of the secondary circuit have developed an open circuit, short circuit, or unwanted grounds.

8.3.3.4 Secondary Voltage Test. When the primary voltage is known, voltage transformers may be tested by measuring the secondary voltage to reveal defects in the transformer or secondary circuit that appreciably affect accuracies.

8.4 Coupling-Capacitor Voltage Transformers

8.4.1 In-Service Performance Tests. A program for periodic test of the transformer should be established to assure maintenance of acceptable accuracy. The frequency of test should be established in accordance with the demonstrated operating stability of the transformer.

9. Pulse Devices

9.1 General

9.1.1 Scope. This section includes the standard requirements, approval tests, and test methods for pulse initiators, relays, and totalizers. Pulse recorders are covered in 5.3.

9.1.2 Acceptable Pulse Devices. New types of pulse devices, in order to be acceptable, shall conform to certain requirements specified in 9.7.1 through 9.7.3, which are intended to determine their reliability and acceptable accuracy insofar as these qualities can be demonstrated by laboratory tests.

9.1.3 Adequacy of Testing Laboratory. Tests for determining the acceptability of the types of pulse devices under these specifications shall be made in a laboratory having adequate facilities, using instruments of an order of accuracy at least equal to that of the shop instruments and standards described in Section 4, Standards and Standardizing Equipment. These instruments should be checked against the laboratory working standards before and after the tests, or more often as required. The tests shall be conducted only by personnel who have thorough practical and theoretical knowledge and adequate training in the making of precision measurements.

9.1.4 Definitions. For definitions, see Section 2.

9.2 Types of Pulse Devices

9.2.1 Basic Type. Pulse initiators, relays, and totalizers are considered to be of the same basic type if they are produced by the same manufacturer, bear a related type designation, are of the same general design, and have the same relationship of parts.

9.2.2 Variations within the Basic Type. Pulse initiators, relays, and totalizers of the same basic type may vary according to the service for which they are designed, such as (but not limited to) the following:

(1) Voltage
(2) Frequency
(3) Type of input (two-wire or three-wire)
(4) Type of output (two-wire or three-wire)
(5) Pulse-initiator ratio
(6) Terminal arrangement

9.2.3 Acceptance of Basic Type in Whole or Part. A basic type of pulse initiator, relay, or totalizer may be accepted as a whole, or a restricted variation of a type may be accepted.

9.2.4 Minor Variations. Minor variations in the mechanical construction, which are not of such character as to affect the operation of the initiator, relay, or totalizer, may be permitted in different pulse initiators, relays, and totalizers of the same basic type.

9.2.5 Pulse Initiators, Relays, and Totalizers Requiring Separate Tests. Pulse initiators, relays, and totalizers of the same basic type but differing in frequency shall be treated as different types for the purposes of approval test.

9.2.6 Special Types. In the case of a type of pulse initiator, relay, or totalizer that comes within the scope of these specifications, but is of such design that some of the tests hereinafter specified are inapplicable or cannot be made under the specified conditions, limited approval may be granted subject to appropriate restrictions.

9.3 Specifications for Design and Construction

9.3.1 Type Designation and Identification. Each pulse initiator, relay, or totalizer shall be designated by type, and may also have a serial number or other marking to identify it. Such identification shall be legibly marked on each pulse initiator, relay, or totalizer.

9.3.2 Sealing. Pulse initiators, relays, and totalizers shall be provided with facilities for sealing to prevent unauthorized entry when mounted external to the watthour meter.

9.3.3 Cover. The cover shall be dustproof and, if intended for outdoor installation, shall also be raintight.

9.3.4 Marking of Leads and Terminals. Leads and terminals of the pulse initiator, relay, or totalizer shall be legibly numbered or identified in such a manner that the identification cannot be obliterated easily.

9.3.5 Construction and Workmanship. Pulse initiators, relays, and totalizers shall be substantially constructed in a workmanlike manner of suitable materials to attain stability of performance and sustained accuracy over long periods of time and over wide ranges of operating conditions with minimum maintenance.

9.3.6 Data to Be Printed on Pulse Devices

9.3.6.1 Pulse Initiator

(1) Type identification
(2) Pulse-initiator output ratio (M_p or R/P) (see Section D2 of Appendix D)
(3) Voltage rating (not applicable to mechanical pulse devices)
(4) Frequency rating (not applicable to mechanical pulse devices)
(5) Type of output (two-wire or three-wire)[16]

9.3.6.2 Pulse Amplifier or Relay

(1) Manufacturer's name or trademark
(2) Type identification
(3) Type of input (two-wire or three-wire)[16]
(4) Type of output (two-wire or three-wire)[16]
(5) Voltage
(6) Frequency
(7) Wiring diagram

---

[16]May be omitted where nameplate space is limited, but shall be included in manufacturer's application data.

9.3.6.3 Totalizing Relay

(1) Manufacturer's name or trademark

(2) Type identification

(3) Input-to-output pulse ratio

(4) Number of additive and subtractive elements. If both are present, each must be clearly identified

(5) Type of input (two-wire or three-wire)[16]

(6) Type of output (two-wire or three-wire)[16]

(7) Voltage

(8) Frequency

9.3.7 Receiving and Transmitting Rate. Pulse initiators, relays, or totalizers shall be capable of receiving pulses or transmitting pulses, or both, continuously at their maximum pulse rate.

9.4 Selection of Pulse Devices for Approval Tests

9.4.1 Samples to Be Representative of Type. The pulse initiators, relays, or totalizers to be tested shall be representative of the type and shall represent the average commercial product of the manufacturer.

9.4.2 Number to Be Tested. A minimum of two devices of a type shall be used to determine the acceptability of the type.

9.5 Conditions of Test

9.5.1 Tests to Be Applied. Each pulse initiator, relay, or totalizer shall be subjected to the tests as specified in 9.7, except that those that are a modification of a type already subjected to the tests or selected for special services may be exempted from certain tests.

9.5.2 Order of Conducting Tests. The items of each test shall be conducted in the order given.

After each test, a sufficient time interval shall be allowed for the pulse device to come to a stable condition before making the next observation or test.

9.5.3 Specific Conditions of Test. The pulse initiators, relays, or totalizers shall be mounted on a support free from vibration.

All tests shall be made at 23 °C ±5 °C, nameplate voltage, and rated frequency, unless otherwise specified.

9.5.4 Initial Condition. The pulse initiators, relays, or totalizers shall be connected so as to initiate or receive pulses, or both, and the pulse output shall be connected to a counter for the purpose of recording the total pulses transmitted.

9.6 Rules Governing the Acceptance of Types

9.6.1 Replacements. Replacements or repairs may be made if physical defects of a minor nature become apparent during the tests. If, during the tests, significant defects in design or manufacture become apparent, the test shall be suspended.

9.6.2 Basis of Acceptable Performance. A pulse initiator, relay, or totalizer shall be considered acceptable under these specifications when all the samples meet the requirements of 9.7.

Table 9.7.3
Performance Test, Pulse Devices

Condition	Test Point in Approximate Percent of Rated Pulse Capacity	Ambient Temperature
Condition (1)	10	23 °C ± 5 °C
Condition (2)	100	23 °C ± 5 °C
Condition (3)	10	50 °C ± 5 °C
Condition (4)	100	50 °C ± 5 °C
Condition (5)	10	−20 °C ± 5 °C
Condition (6)	100	−20 °C ± 5 °C

9.7 Performance Requirements

9.7.1 Mechanical Load. The mechanical load imposed on the meter by the pulse initiator shall be within the adjustment range of the meter. This load shall be as nearly constant as practical throughout the entire cycle of operation of the pulse initiator.

9.7.2 Insulation. The voltage-withstand (dielectric) test shall consist of applying a 60-Hz voltage of 1.5 kV rms for 1 minute between current-carrying parts and frame.

NOTE: Low-voltage electronic circuits and the pulse terminals (KYZ) are not to be subjected to the insulation test.

9.7.3 Performance Test. The pulse device shall be placed in a space having a temperature of 23 °C ±5 °C, and allowed to stand for not less than 2 hours with the voltage circuits energized at approximately nameplate voltage. This operation shall be repeated at the various values of temperature shown in Table 9.7.3.

The pulse device shall operate at each test condition for at least 1 hour. An acceptable pulse device shall not gain but may lose one pulse when all pulse circuits are energized under any of the test conditions in Table 9.7.3 at 85%, 100%, and 110% of nameplate voltage.

9.8 Installation and In-Service Requirements

9.8.1 Installation. The installation requirements as outlined for meters in Section 7 shall be applicable to pulse devices.

9.8.2 In-Service Conditions. All connections shall be in such condition as to provide good electrical contact.

Terminal markings shall be legible.

There shall be no evidence of physical damage to any mechanical parts or any wiring.

There shall be no evidence of thermal overload on the insulation, contacts, terminals, or other component parts.

9.8.3 In-Service Performance. The utilization of pulse devices in a metering system shall not add errors to the billing registration so that overall errors exceed those given in 8.2.

10. Auxiliary Devices for Electricity Meters

10.1 General

10.1.1 Scope. This section includes the standard requirements, approval tests, and test methods for auxiliary devices that are commonly used with electricity meters. Included in this section are:

(1) Phase-shifting transformers

(2) Transformer-loss compensators

10.1.2 Acceptable Auxiliary Devices. New types of auxiliary devices, in order to be acceptable, shall conform to certain requirements specified in 10.2.7 and 10.3.7, which are intended to determine their reliability and acceptable accuracy insofar as these qualities can be demonstrated by laboratory tests.

10.1.3 Adequacy of Testing Laboratory. Tests for determining the acceptability of the types of auxiliary devices under these specifications shall be made in a laboratory having adequate facilities, using instruments of an order of accuracy at least equal to that of the shop instruments and standards described in Section 4. These instruments should be checked against the laboratory working standards before and after the tests, or more often as required. The tests shall be conducted only by personnel who have thorough practical and theoretical knowledge and adequate training in the making of precision measurements.

10.2 Phase-Shifting Transformers

10.2.1 Definition. For definition of a phase-shifting transformer, see Section 2.

10.2.2 Types Defined. Phase-shifting transformers are considered to be of the same type if they are produced by the same manufacturer, bear the manufacturer's same type designation, are of the same general design, and have the same relationship of parts. They shall have the same phasor diagram and be substantially equivalent in the following respects to be grouped as one type for approval tests:

(1) Rated volts per turn

(2) Length, cross section, and shape of magnetic circuit

(3) Characteristics of the core material

(4) Arrangement of the coils with respect to the magnetic circuit

(5) Relation of resistance of the windings to the rated voltage

10.2.3 Specifications for Design and Construction

10.2.3.1 Type Designation and Identification. Each phase-shifting transformer shall be designated by type, and given a serial number by the manufacturer. The serial number and type designation shall be legibly marked on the nameplate of each transformer.

10.2.3.2 Marking of Tap Leads and Terminals. Terminals (taps) of phase-shifting transformers shall be legibly numbered in such a manner that the numbers cannot be easily obliterated. The numbers shall correspond to those indicated on a suitable phasor diagram, showing the percentage of applied voltage and the phase angle between the input and output voltages for all taps.

10.2.3.3 Construction and Workmanship. The phase-shifting transformers shall be substantially constructed in a workmanlike manner of suitable material to attain stability of performance and sustained accuracy over long periods of time and over wide ranges of operating conditions with minimum maintenance. The cover shall be sealable, fastenings shall be accessible, and the cover shall be arranged so that it can be removed for inspection purposes. The phase-shifting transformer shall be provided with suitable mounting facilities so that it can be rigidly fastened in place. Such mounting facilities may consist of holes in the frame to accommodate bolts or screws, or the transformer may be rigidly fastened in a container having suitable mounting facilities supplied as part of the container. Provision shall be made for externally grounding the transformer cores, if accessible.

10.2.3.4 Nameplates. Phase-shifting transformers shall be provided with nameplates that shall include, as a minimum, the following information:

(1) Manufacturer's name or trademark
(2) Manufacturer's serial number
(3) Manufacturer's type
(4) Voltage ratings
(5) Type of service (for example, three-phase four-wire wye)
(6) Frequency

10.2.4 Selection of Phase-Shifting Transformers for Approval Test

10.2.4.1 Samples to Be Representative of Type. The transformers to be tested shall be representative of the type and shall represent the average commercial product of the manufacturer.

10.2.4.2 Number to Be Tested. A minimum of two transformers shall be tested to determine the acceptability of the type. When the specimens representing a given type include different voltage ratings or are for different types of service or for different frequencies, there shall be not less than two transformers of each of the representative voltage ratings or types of service.

10.2.5 Conditions of Test

10.2.5.1 Alternating-Current Tests. All alternating-current tests shall be conducted on a circuit supplied by a sine-wave source with a distortion factor not greater than 3%.

10.2.5.2 Order of Conducting Tests. The items of each test shall be conducted in the order given.

After each change in voltage or load, a sufficient time interval shall be allowed for the phase-shifting transformer to come to a stable condition before making the next observation or test.

10.2.6 Basis of Acceptable Performance. A phase-shifting transformer shall be considered acceptable under these specifications when all the samples meet the requirements of 10.2.7 and the tap leads and terminals are marked in accordance with the manufacturer's diagram.

10.2.7 Performance Requirements

10.2.7.1 Insulation. The voltage-withstand (dielectric) test shall consist of applying a 60-Hz voltage of 2.5 kV rms for 1 minute be-

159

tween the windings and core, with the latter connected to the grounded case.

10.2.7.2 Test No 1: Performance. Accuracy tests shall be performed according to the methods described in 10.2.7.2.1 and 10.2.7.2.2.

10.2.7.2.1 Polyphase Test. With balanced polyphase rated voltage applied to the input terminals and burdens of 12.5 volt-amperes at 0.1 power factor connected to the corresponding output terminals for the specified phase shift, all input and output voltages shall be measured. Phase-shifting transformers having more than one burden rating shall be tested at each rated burden. The allowable tolerance of the output voltage shall be within ± 1.0% in terms of the input voltage.

10.2.7.2.2 Single-Phase Test. With approximate rated voltage applied to the input terminals and without any burdens connected to the tap terminals, all tap voltages shall be measured. The magnitudes of the measured voltages, converted to percent of input voltage, shall agree within ± 1.0% of the theoretical values given in the manufacturer's published data, which include appropriate wiring and phasor diagrams.

10.2.7.3 Test No 2: Temperature Rise. The temperature-rise test shall be conducted at maximum rated voltage and at burdens of 12.5 voltamperes at 0.1 power factor or at the maximum rated burdens. This test shall be conducted in the manner outlined for instrument transformers in 6.3.4. The values so obtained shall not exceed the allowable temperature rise for the class of insulation employed.

10.2.8 Installation and In-Service Requirements

10.2.8.1 Installation. The general requirements for installation of meters as outlined in Section 7 shall normally apply for phase-shifting transformers.

10.2.8.2 In-Service Conditions. Terminal connections shall be in such condition as to provide good electrical contact.

Terminal numbers shall be clearly visible.

There shall be no evidence of physical damage to the switches or internal wiring.

There shall be no evidence of thermal overload on the insulation, switches, terminals, etc.

10.2.8.3 In-Service Performance. The output voltage values in terms of percentage of input voltage shall be within ± 2.0% when tested under the conditions of 10.2.7.2.

Phase-shifting transformers returned to the laboratory shall meet the original specifications, except that the voltage-withstand test shall be made at 1.5 kV.

10.3 Transformer-Loss Compensators

10.3.1 Definition. For definition of transformer-loss compensators, see Section 2.

10.3.2 Types Defined

10.3.2.1 Definition of Type. Transformer-loss compensators are considered to be of the same type if they are produced by the same manufacturer, bear the manufacturer's same type designation, are of the same general design, and have the same relationship of parts.

10.3.2.2 Variations within the Type. Transformer-loss compensators are considered to be of three general types:

(1) Single-element compensators are intended for use with a single-phase meter, or for connection to one stator of a multistator meter for use on a *balanced* polyphase circuit.

(2) Two-element compensators are intended for use with a two-stator meter on balanced or unbalanced loads with any type of service for which the meter may be intended.

(3) Three-element compensators are intended for use with a three-stator meter connected to balanced or unbalanced loads.

10.3.3 Specifications for Design and Construction

10.3.3.1 Type Designation and Identification. Each transformer-loss compensator shall be designated by type and given a serial number by the manufacturer. The serial number and type designation shall be legibly marked on the nameplate of each transformer-loss compensator.

10.3.3.2 Construction and Workmanship. The transformer-loss compensator shall be substantially constructed in a workmanlike manner of suitable materials to attain stability of performance and sustained accuracy over long periods of time and over wide ranges of operating conditions with minimum maintenance. The transformer-loss compensator shall be provided with suitable mounting facilities so that it can be rigidly fastened in place.

10.3.3.3 Marking of Terminals. Terminals of transformer-loss compensators shall be legibly marked to identify current and voltage circuits in such a manner that the designations shall correspond to the manufacturer's standard connection diagrams.

10.3.3.4 Nameplate. Transformer-loss compensators shall be provided with nameplates that shall include, as a minimum, the following information:

(1) Manufacturer's name or trademark
(2) Manufacturer's serial number
(3) Number of meter stators for which the compensator is intended
(4) Voltage
(5) Frequency

10.3.3.5 Cover. The cover shall be sealable, and the fastenings shall be accessible.

10.3.4 Selection of Transformer-Loss Compensators for Approval Test

10.3.4.1 Samples to Be Representative of Type. Any transformer-loss compensator submitted for test shall be representative of its type and shall represent the average commercial product of the manufacturer.

10.3.4.2 Number to Be Tested. A minimum of two devices of a type shall be tested to determine the acceptability of the type.

10.3.5 Conditions of Test

10.3.5.1 Alternating-Current Source. All alternating-current tests shall be conducted on a circuit supplied by a sine-wave source with a distortion factor not greater than 3%.

10.3.5.2 Order of Conducting Tests. The tests on transformer-loss compensators shall be conducted in the order outlined in 10.3.7.

10.3.6 Basis of Acceptable Performance. A transformer-loss compensator shall be considered acceptable under these specifications when each of the following requirements is satisfied:

(1) It shall pass the voltage-withstand (dielectric) tests as specified in 10.3.7.1.

(2) The percentage registration of load-plus-loss, as related to the desired correct values, shall be within the limits given in 10.3.7 for all tests.

(3) Terminals are marked in accordance with the manufacturer's diagram.

10.3.7 Performance Requirements

10.3.7.1 Insulation. The voltage-withstand (dielectric) test shall consist of applying a 60 Hz voltage of 2.5 kV rms between all windings of the transformer-loss compensator and a grounded metal plate on which the compensator is mounted.

10.3.7.2 Preliminary Operations. The compensator shall be connected to a suitable meter that has been calibrated to read correctly at light load, heavy load, and inductive load, and if the meter is of a multistator type, its stator balance shall be within acceptable limits as specified in 5.1.8.1. Throughout the tests, no further adjustments shall be made on the meter.

The compensator shall be adjusted to neutralize the effects of the voltage-circuit current of the meter stators in accordance with the manufacturer's instructions. The same meter shall be used with the compensator throughout the tests.

The compensator shall be adjusted for an iron loss in watts equal to 1% of the secondary meter rating. For purposes of these tests, meter rating is defined as the product of the rated voltage of the meter, the secondary rating of current transformers for which the meter is designed, and the number of elements of the compensator (that is, for 120-volt meters, 600 voltamperes for a single-stator meter, 1200 voltamperes for a two-stator meter, and 1800 voltamperes for a three-stator meter). As the iron-loss adjustment is made at 10% of rated current, correct adjustment of the compensator for this loss will cause the meter to operate at 110% registration.

The copper-loss elements of the compensator shall be adjusted for 5% copper loss at the meter rating on each element of the compensator. Throughout the remainder of the tests, further adjustments shall not be made on the compensator.

The desired performance in the following tests is based on compensation for copper loss in accordance with the square of the current and on variation of iron loss in accordance with the square of the voltage.

If the transformer-loss compensators under test are designed for use with multistator meters, the test described in 10.3.7.3 may be made with the meter stators and loss-compensator elements con-

Table 10.3.7.3
Effect of Variation of Current,
Transformer-Loss Compensators

Condition	Percent Rated Current	Percent Registration, Meter without Compensator*	Desired Percent Registration, Meter with Compensator	Maximum Deviation in Percent from Desired Registration
Condition (1)	10	100	100.5	± 0.3
Condition (2)	20	100	101.0	± 0.3
Condition (3)	50	100	102.5	± 0.3
Condition (4)	100	100	105.0	± 0.3
Condition (5)	150	100	107.5	± 0.3
Condition (6)	200	100	110.0	± 0.3

*Although the meter should not be adjusted for each reading, correction for any deviation from 100.0% for the meter without compensator should be made in determining the deviation from the desired performance.

nected in series, in accordance with the manufacturer's standard connections.

10.3.7.3 Test No 1: Effect of Variation of Current on Copper-Loss Registration. The test shall be made at nameplate voltage, rated frequency, and with 1.0 power factor load. The iron-loss compensation shall be disconnected for this test.

The effect of copper-loss registration shall not deviate from the desired performance of the meter with compensator by an amount exceeding that specified in Table 10.3.7.3.

Table 10.3.7.4

Effect of Variation of Load at Unity Power Factor, Transformer-Loss Compensators

Condition	Percent Rated Current	Percent Registration, Meter without Compensator*	Desired Percent Registration, Meter with Compensator	Maximum Deviation in Percent from Desired Registration
Condition (1)	10	100	110.5	± 0.3
Condition (2)	20	100	106.0	± 0.3
Condition (3)	50	100	104.5	± 0.3
Condition (4)	100	100	106.0	± 0.3
Condition (5)	150	100	108.2	± 0.3
Condition (6)	200	100	110.5	± 0.3

*Although the meter should not be adjusted for each reading, correction for any deviation from 100.0% for the meter without compensator should be made in determining the deviation from the desired performance.

10.3.7.4 Test No 2: Effect of Variation of Load on Total-Loss Registration. The test shall be made at calibration voltage, rated frequency, and with unity power factor load. Both iron-loss and copper-loss compensations shall be included in this and the following tests.

The effect of loss registration shall not deviate from the desired performance of the meter with compensator by an amount exceeding that specified in Table 10.3.7.4.

Table 10.3.7.5
Effect of Variation of Load at Lagging Power Factor,
Transformer-Loss Compensators

Condition	Percent Rated Current	Percent Registration, Meter without Compensator*	Desired Percent Registration, Meter with Compensator	Maximum Deviation in Percent from Desired Registration
Condition (1)	10	100	121.0	±0.3
Condition (2)	20	100	112.0	±0.3
Condition (3)	50	100	109.0	±0.3
Condition (4)	100	100	112.0	±0.3
Condition (5)	150	100	116.3	±0.3
Condition (6)	200	100	121.0	±0.3

*Although the meter should not be adjusted for each reading, correction for any deviation from 100.0% for the meter without compensator should be made in determining the deviation from the desired performance.

10.3.7.5 Test No 3: Effect of Variation of Load on Loss Registration at Lagging Power Factor. The test shall be made at 0.5 power factor lag. Both iron-loss and copper-loss compensations shall be included.

NOTE: In testing loss compensators it is important that the power factor be exact. This is because the *desired performance* of the compensator varies with the power factor under which the test is made.

The effect of loss registration shall not deviate from the desired performance of the meter with compensator by an amount exceeding that specified in Table 10.3.7.5.

Table 10.3.7.6
Effect of Variation of Voltage,
Transformer-Loss Compensators

Condition	Percent Calibration Voltage	Percent Rated Current	Percent Registration, Meter without Compensator*	Desired Percent Registration, Meter with Compensator	Maximum Deviation in Percent from Desired Registration
Condition (1)	100	10	100	110.5	± 0.3
Condition (2)	110	10	100	111.5	± 0.3
Condition (3)	90	10	100	109.5	± 0.3
Condition (4)	100	100	100	106.0	± 0.3
Condition (5)	110	100	100	106.1	± 0.3
Condition (6)	90	100	100	105.9	± 0.3

*Although the meter should not be adjusted for each reading, correction for any deviation from 100.0% for the meter without compensator should be made in determining the deviation from the desired performance.

10.3.7.6 Test No 4: Effect of Variation of Voltage on Loss Registration. The test shall be made at 10% and 100% of rated current, rated frequency, and with 1.0 power factor load.

The effect of loss registration shall not deviate from the desired performance of the meter with compensator by an amount exceeding that specified in Table 10.3.7.6.

10.3.7.7 Test No 5: Temperature Rise. The temperature-rise test shall be conducted at rated current, nameplate voltage, and 1.0 power factor, with the compensator adjusted in accordance with 10.3.7.2. The temperature rise shall be determined by measurement after the compensator has been in continuous operation for 6 hours. The values so obtained shall not exceed the allowable temperature rise for the class of insulation employed in the various parts.

10.3.8 Installation and In-Service Requirements

10.3.8.1. Installation. The general requirements for installation of meters as outlined in Section 7 shall normally apply for transformer-loss compensators. They shall be so installed as to facilitate in-service inspection, test, and calibration.

10.3.8.2 In-Service Conditions. Terminal connections shall be in such condition as to provide good electrical contact. Terminal designations shall be clearly visible.

There shall be no evidence of physical damage to the component parts, their adjustments, or to the internal wiring.

There shall be no evidence of thermal overload on the insulation, resistors, terminals, etc.

10.3.8.3 In-Service Performance. Transformer-loss compensators shall be tested on the same schedule and at the same time as the meters with which they are associated.

In-service tests on transformer-loss compensators shall be made at the normal service test points of the meter. Performance deviations determined in accordance with 10.3.7 as related to the desired performance for the installation shall not exceed ± 0.3%.

Transformer-loss compensators returned to the laboratory shall meet the original specifications except that the voltage-withstand test shall be made at 1.5 kV rms.

11. References

When the following standards referred to in this standard are superseded by an approved revision, the revision shall apply.

[1] ANSI C39.1-1981, Requirements for Electrical Analog Indicating Instruments.

[2] ANSI C39.5-1974, Safety Requirements for Electrical and Electronic Measuring and Controlling Instrumentation.

[3] ANSI C39.6-1969 (R1975), Requirements for Automatic Digital Voltmeters and Ratio Meters.

[4] ANSI/IEEE C57.13-1978, IEEE Standard Requirements for Instrument Transformers.

[5] ANSI/IEEE Std 4-1978, IEEE Standard Techniques for High-Voltage Testing.

[6] ANSI/IEEE Std 100-1977, IEEE Standard Dictionary of Electrical and Electronics Terms.

[7] ANSI/NFPA 70-1981, National Electrical Code.

[8] IEEE Std 1-1969, IEEE Standard General Principles for Temperature Limits in the Rating of Electric Equipment.

Appendixes

(These Appendixes are not a part of ANSI C12.1-1982, American National Standard Code for Electricity Metering.)

Appendix A
Instrument Transformer Parallelograms

A1. Current Transformers

The relationship betweeen the limits of ratio correction factors and phase angles for the limiting values of transformer correction factors for current transformers in Table 40 of this standard is shown by parallelograms in Figs A1, A2, and A3.

The ratio correction factor and phase angle of the current transformer shall be within parallelogram A at 100% current and at maximum continuous current, and within parallelogram B at 10% current for the indicated accuracy class.

Fig A1
Limits for Accuracy Class 0.3
for Current Transformers for Metering Service

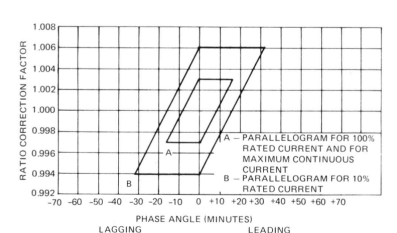

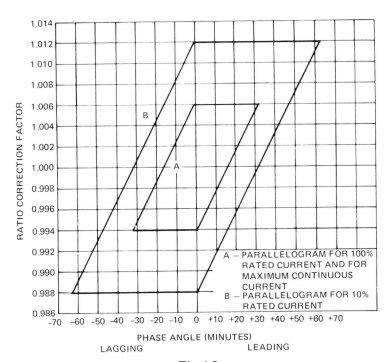

Fig A2
Limits for Accuracy Class 0.6
for Current Transformers for Metering Service

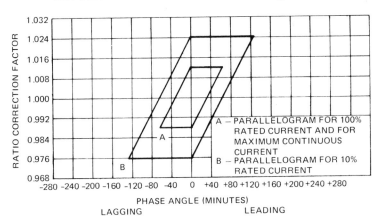

Fig A3
Limits for Accuracy Class 1.2
for Current Transformers for Metering Service

169

A2. Voltage Transformers

The relationship between the limits of ratio correction factors and phase angles for the limiting values of transformer correction factors for voltage transformers in Table 5.4.6.2 of this standard is shown by parallelograms in Fig A4.

The ratio correction factor and phase angle of the voltage transformer shall be within the parallelogram for the indicated accuracy class. These limits shall apply from 10% below to 10% above rated primary voltage at rated frequency, and from zero burden on the voltage transformer to the rated burden.

Fig A4
Limits for Accuracy Classes 0.3, 0.6, and 1.2
for Voltage Transformers for Metering Service

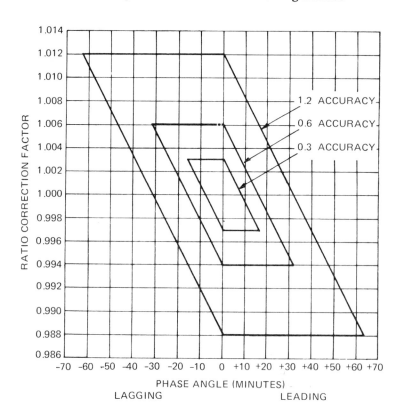

Appendix B
Determination of Voltage Transformer Accuracy by the Circle Method

The circle method provides an easy means for determining the accuracy of a voltage transformer at any desired burden by using only the phase angle and ratio correction factor (RCF) at a zero volt-ampere (VA) burden and one other known burden. Normally the manufacturer furnishes this information with the transformer.

Graph paper scaled such that a given distance represents 0.0010 unit on the RCF axis and 3.438 minutes on the phase-angle axis is necessary, as indicated in Fig B1.

Fig B1
Sample of Graph Paper Specifically Scaled for Voltage Transformer Circle Diagram

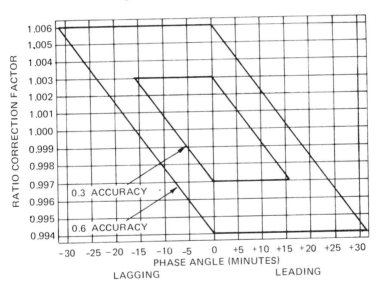

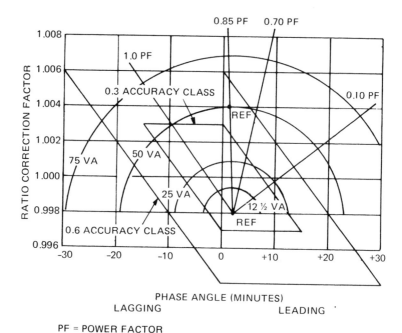

Fig B2
Example of Application of Circle Diagram

EXAMPLE: The following data are test results at 120 secondary volts for a 2400:120-volt voltage transformer. At 0 VA burden: RCF = 0.9979; phase angle = +2.0 minutes. At 50 VA, 85% power factor burden: RCF = 1.0040; phase angle = +1.0 minute. Reference points representing the two sets of given performance data are plotted on the grid as shown in Fig B2. Points on the line connecting the two given burdens represent performances at various burdens with the same power factor. Performance at other volt-ampere and power-factor burdens can now be plotted by making radii proportional to voltamperes and angles equal to burden-power-factor angles.

Appendix C
Application of Instrument Transformer Correction Factors
C1. General

As stated in 6.1 of this standard, transformer errors may be neglected in the calibration of the meter if instrument transformers are used that conform to the 0.3-accuracy-class limits with the actual secondary burden.

Instrument transformer test cards show the ratio correction factor (RCF) and the phase angle of the transformer in minutes, positive or negative. When the ratio correction factor and phase angle of an instrument transformer, or set of instrument transformers, to be used with a watthour meter are known, a determination of the required meter adjustment may be made either by use of the nomograph in Fig C1 or by use of the formulas or data in Tables C1 and C2.

C2. Nomograph

C2.1 Current Transformers. For a transformer-rated meter with current transformers only, the average transformer corrections may be obtained as follows:

(1) Calculate the average RCF of all current transformers connected to the meter for heavy load and for light load.

EXAMPLE 1:

	Ratio Correction Factor	
	Heavy Load	Light Load
CT No 1	1.002	1.008
CT No 2	0.997	1.006
CT No 3	0.998	0.998
Average RCF	0.999	1.004

(2) Calculate the algebraic average phase angle for heavy load and for light load.

EXAMPLE 2:

	Phase Angle	
	Heavy Load	Light Load
CT No 1	+ 1'	+25'
CT No 2	−13'	− 3'
CT No 3	−15'	− 4'
Average phase angle	− 9'	+ 6'

(3) Obtain the heavy-load and light-load average transformer corrections by use of the nomograph in Fig C1, as follows:

(a) Heavy Load. Lay a straightedge across the chart so that it intersects the extreme left-hand vertical scale at the average RCF value (0.999) and intersects the extreme right-hand vertical scale at the figure representing the average phase-angle value (−9'). The correction to be applied to the meter is the value obtained where this line

173

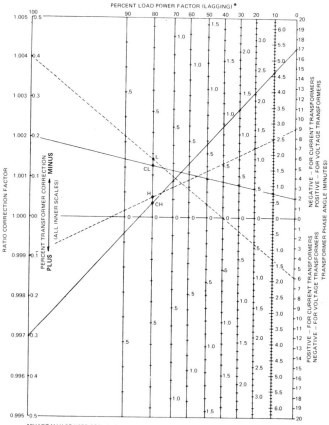

Fig C1

Instrument Transformer Correction Chart

PERCENT LOAD POWER FACTOR (LAGGING) *

*CHART MAY BE USED FOR LEADING LOAD POWER FACTORS PROVIDED THE ALGEBRAIC SIGNS
OF THE CURRENT AND VOLTAGE TRANSFORMER PHASE ANGLES ARE CONSIDERED AS REVERSED.

Table C1
Correction Factors (cos θ/cos θ_2)* for Phase Angle, for Lagging Current When ($\beta - \gamma$) Is Positive or for Leading Current When ($\beta - \gamma$) Is Negative

Phase Angle ($\beta - \gamma$)	APPARENT POWER FACTOR (cos θ_2)													
	0.10	0.15	0.20	0.25	0.30	0.40	0.50	0.60	0.70	0.80	0.90	0.95	0.99	1.00
5′	0.9855	0.9904	0.9929	0.9944	0.9954	0.9967	0.9975	0.9981	0.9985	0.9989	0.9993	0.9995	0.9998	1.0000
10′	0.9711	0.9808	0.9857	0.9887	0.9907	0.9933	0.9950	0.9961	0.9970	0.9978	0.9986	0.9990	0.9996	1.0000
15′	0.9566	0.9712	0.9786	0.9831	0.9861	0.9900	0.9924	0.9942	0.9955	0.9967	0.9979	0.9986	0.9994	1.0000
20′	0.9421	0.9616	0.9715	0.9775	0.9815	0.9867	0.9899	0.9922	0.9940	0.9956	0.9972	0.9981	0.9992	1.0000
25′	0.9276	0.9520	0.9643	0.9718	0.9768	0.9833	0.9874	0.9903	0.9926	0.9945	0.9965	0.9976	0.9980	1.0000
30′	0.9131	0.9424	0.9572	0.9662	0.9722	0.9800	0.9848	0.9883	0.9911	0.9934	0.9957	0.9971	0.9987	1.0000
40′	0.8842	0.9232	0.9429	0.9549	0.9629	0.9733	0.9798	0.9844	0.9881	0.9912	0.9943	0.9961	0.9983	0.9999
50′	0.8552	0.9040	0.9286	0.9436	0.9536	0.9666	0.9747	0.9805	0.9851	0.9900	0.9929	0.9951	0.9978	0.9999
1° 0′	0.8262	0.8848	0.9143	0.9323	0.9444	0.9599	0.9696	0.9766	0.9820	0.9868	0.9914	0.9941	0.9974	0.9998
10′	0.7972	0.8656	0.9000	0.9209	0.9350	0.9531	0.9645	0.9726	0.9790	0.9845	0.9899	0.9931	0.9969	0.9998
20′	0.7682	0.8464	0.8857	0.9096	0.9257	0.9464	0.9594	0.9687	0.9760	0.9823	0.9885	0.9921	0.9964	0.9997
30′	0.7393	0.8271	0.8714	0.8983	0.9164	0.9397	0.9543	0.9648	0.9730	0.9800	0.9870	0.9911	0.9959	0.9997
40′	0.7102	0.8079	0.8571	0.8869	0.9071	0.9329	0.9492	0.9608	0.9699	0.9778	0.9855	0.9900	0.9954	0.9996
50′	0.6812	0.7886	0.8428	0.8750	0.8978	0.9262	0.9441	0.9568	0.9668	0.9755	0.9840	0.9890	0.9949	0.9995
2° 0′	0.6521	0.7694	0.8234	0.8642	0.8884	0.9194	0.9389	0.9529	0.9638	0.9732	0.9825	0.9879	0.9944	0.9994
10′	0.6231	0.7501	0.8141	0.8529	0.8791	0.9127	0.9338	0.9490	0.9607	0.9709	0.9810	0.9869	0.9939	0.9993
20′	0.5941	0.7308	0.7997	0.8415	0.8697	0.9059	0.9287	0.9449	0.9576	0.9686	0.9795	0.9853	0.9934	0.9992
30′	0.5650	0.7115	0.7854	0.8301	0.8603	0.8991	0.9235	0.9409	0.9545	0.9663	0.9779	0.9847	0.9928	0.9990
40′	0.5360	0.6923	0.7710	0.8187	0.8510	0.8923	0.9183	0.9369	0.9515	0.9640	0.9764	0.9836	0.9923	0.9989
50′	0.5069	0.6730	0.7566	0.8073	0.8416	0.8855	0.9132	0.9329	0.9483	0.9617	0.9748	0.9825	0.9917	0.9988
3° 0′	0.4779	0.6537	0.7422	0.7959	0.8322	0.8787	0.9080	0.9288	0.9452	0.9594	0.9733	0.9814	0.9912	0.9986
10′	0.4488	0.6344	0.7279	0.7845	0.8228	0.8719	0.9028	0.9248	0.9421	0.9570	0.9717	0.9803	0.9906	0.9985
20′	0.4198	0.6151	0.7135	0.7731	0.8134	0.8651	0.8976	0.9208	0.9390	0.9547	0.9701	0.9792	0.9900	0.9983
30′	0.3907	0.5957	0.6991	0.7617	0.8040	0.8582	0.8924	0.9167	0.9359	0.9523	0.9686	0.9781	0.9894	0.9981
40′	0.3616	0.5764	0.6847	0.7503	0.7946	0.8514	0.8872	0.9127	0.9327	0.9500	0.9670	0.9769	0.9888	0.9980
50′	0.3326	0.5571	0.6702	0.7388	0.7852	0.8446	0.8820	0.9086	0.9296	0.9476	0.9654	0.9758	0.9882	0.9978
4° 0′	0.3035	0.5378	0.6558	0.7274	0.7758	0.8377	0.8767	0.9046	0.9264	0.9452	0.9638	0.9746	0.9876	0.9976
10′	0.2744	0.5185	0.6414	0.7160	0.7663	0.8309	0.8715	0.9005	0.9232	0.9429	0.9622	0.9735	0.9870	0.9974
20′	0.2453	0.4901	0.6270	0.7045	0.7569	0.8240	0.8663	0.8964	0.9201	0.9405	0.9605	0.9723	0.9864	0.9971
30′	0.2163	0.4798	0.6125	0.6930	0.7474	0.8171	0.8610	0.8923	0.9169	0.9381	0.9589	0.9711	0.9857	0.9969
40′	0.1872	0.4604	0.5981	0.6816	0.7380	0.8103	0.8558	0.8882	0.9137	0.9357	0.9573	0.9699	0.9851	0.9967
50′	0.1581	0.4411	0.5837	0.6701	0.7285	0.8034	0.8505	0.8841	0.9105	0.9333	0.9556	0.9687	0.9844	0.9964
5° 0′	0.1290	0.4217	0.5692	0.6586	0.7191	0.7965	0.8452	0.8800	0.9073	0.9308	0.9540	0.9675	0.9838	0.9962
10′	0.0999	0.4024	0.5548	0.6472	0.7096	0.7896	0.8400	0.8759	0.9041	0.9284	0.9523	0.9663	0.9831	0.9959
20′	0.0708	0.3830	0.5403	0.6357	0.7001	0.7827	0.8347	0.8717	0.9008	0.9260	0.9507	0.9651	0.9824	0.9957

* $\dfrac{\cos \theta}{\cos \theta_2}$ may be also written $\dfrac{\cos (\theta_2 + \beta - \gamma)}{\cos \theta_2}$

Interpolation for correction factors corresponding to values of ($\beta - \gamma$) lying between those given in the table may be made without error. Interpolation for correction factors corresponding to values of cos θ_2 lying between those given in the table may be made without exceeding an error of 0.0010 in the sections of the table lying between the heavy black lines; outside of these sections, and in all cases where the adjacent values of cos θ_2 are separated by the heavy black lines, the maximum error in interpolation will exceed 0.0010.

Table C2
Correction Factors (cos θ/cos θ_2)* for Phase Angle, for Lagging Current When ($\beta - \gamma$) Is Negative or for Leading Current When ($\beta - \gamma$) Is Positive

Phase Angle ($\beta - \gamma$)	APPARENT POWER FACTOR (cos θ_2)													
	0.10	0.15	0.20	0.25	0.30	0.40	0.50	0.60	0.70	0.80	0.90	0.95	0.99	1.00
5'	1.0145	1.0096	1.0071	1.0056	1.0046	1.0033	1.0025	1.0019	1.0015	1.0011	1.0007	1.0005	1.0002	1.0000
10'	1.0289	1.0192	1.0142	1.0113	1.0092	1.0067	1.0050	1.0039	1.0030	1.0022	1.0014	1.0010	1.0004	1.0000
15'	1.0434	1.0288	1.0214	1.0169	1.0139	1.0100	1.0075	1.0058	1.0044	1.0033	1.0021	1.0014	1.0006	1.0000
20'	1.0579	1.0383	1.0285	1.0225	1.0185	1.0133	1.0101	1.0077	1.0059	1.0043	1.0028	1.0019	1.0008	1.0000
25'	1.0723	1.0479	1.0356	1.0281	1.0231	1.0166	1.0126	1.0097	1.0074	1.0054	1.0035	1.0024	1.0010	1.0000
30'	1.0868	1.0575	1.0427	1.0338	1.0277	1.0200	1.0151	1.0116	1.0089	1.0065	1.0042	1.0028	1.0012	1.0000
40'	1.1157	1.0766	1.0569	1.0450	1.0369	1.0266	1.0201	1.0154	1.0118	1.0087	1.0056	1.0038	1.0016	0.9999
50'	1.1446	1.0958	1.0711	1.0562	1.0461	1.0332	1.0251	1.0193	1.0147	1.0108	1.0069	1.0047	1.0020	0.9999
1° 0'	1.1735	1.1149	1.0853	1.0674	1.0553	1.0398	1.0301	1.0231	1.0177	1.0129	1.0083	1.0056	1.0023	0.9998
10'	1.2024	1.1340	1.0995	1.0787	1.0645	1.0464	1.0351	1.0269	1.0206	1.0151	1.0097	1.0065	1.0027	0.9998
20'	1.2313	1.1531	1.1137	1.0898	1.0737	1.0530	1.0400	1.0308	1.0235	1.0172	1.0110	1.0074	1.0030	0.9997
30'	1.2601	1.1722	1.1279	1.1010	1.0829	1.0596	1.0450	1.0346	1.0264	1.0193	1.0123	1.0083	1.0034	0.9997
40'	1.2890	1.1913	1.1421	1.1122	1.0921	1.0662	1.0500	1.0384	1.0292	1.0214	1.0137	1.0091	1.0037	0.9996
50'	1.3178	1.2104	1.1562	1.1234	1.1012	1.0728	1.0549	1.0421	1.0321	1.0235	1.0150	1.0100	1.0040	0.9995
2° 0'	1.3466	1.2294	1.1704	1.1346	1.1104	1.0794	1.0598	1.0459	1.0350	1.0256	1.0163	1.0109	1.0044	0.9994
10'	1.3755	1.2485	1.1845	1.1457	1.1195	1.0859	1.0648	1.0497	1.0379	1.0276	1.0176	1.0117	1.0047	0.9993
20'	1.4043	1.2675	1.1986	1.1569	1.1286	1.0925	1.0697	1.0535	1.0407	1.0297	1.0189	1.0126	1.0050	0.9992
30'	1.4331	1.2866	1.2127	1.1680	1.1377	1.0990	1.0746	1.0572	1.0435	1.0318	1.0202	1.0134	1.0053	0.9990
40'	1.4618	1.3056	1.2268	1.1791	1.1469	1.1055	1.0795	1.0610	1.0464	1.0338	1.0215	1.0142	1.0055	0.9990
50'	1.4906	1.3246	1.2409	1.1902	1.1560	1.1120	1.0844	1.0647	1.0492	1.0359	1.0227	1.0150	1.0058	0.9988
3° 0'	1.5194	1.3436	1.2550	1.2013	1.1650	1.1185	1.0893	1.0684	1.0520	1.0379	1.0240	1.0158	1.0061	0.9986
10'	1.5481	1.3626	1.2691	1.2124	1.1741	1.1250	1.0942	1.0721	1.0548	1.0399	1.0252	1.0166	1.0063	0.9985
20'	1.5768	1.3816	1.2832	1.2235	1.1832	1.1315	1.0990	1.0758	1.0576	1.0419	1.0265	1.0174	1.0066	0.9983
30'	1.6056	1.4005	1.2972	1.2346	1.1923	1.1380	1.1039	1.0795	1.0604	1.0439	1.0277	1.0182	1.0068	0.9981
40'	1.6343	1.4195	1.3113	1.2456	1.2013	1.1445	1.1087	1.0832	1.0632	1.0459	1.0289	1.0190	1.0071	0.9980
50'	1.6630	1.4384	1.3253	1.2567	1.2103	1.1509	1.1136	1.0869	1.0660	1.0479	1.0301	1.0197	1.0073	0.9978
4° 0'	1.6916	1.4573	1.3393	1.2677	1.2194	1.1574	1.1184	1.0906	1.0687	1.0499	1.0313	1.0205	1.0075	0.9976
10'	1.7203	1.4763	1.3533	1.2788	1.2284	1.1638	1.1232	1.0942	1.0715	1.0519	1.0325	1.0212	1.0077	0.9974
20'	1.7489	1.4952	1.3673	1.2898	1.2374	1.1703	1.1280	1.0979	1.0742	1.0538	1.0337	1.0220	1.0079	0.9971
30'	1.7776	1.5141	1.3813	1.3008	1.2464	1.1767	1.1328	1.1015	1.0770	1.0558	1.0349	1.0227	1.0081	0.9969
40'	1.8062	1.5329	1.3953	1.3118	1.2554	1.1831	1.1376	1.1052	1.0797	1.0577	1.0361	1.0234	1.0083	0.9967
50'	1.8348	1.5518	1.4092	1.3228	1.2644	1.1895	1.1424	1.1088	1.0824	1.0596	1.0373	1.0241	1.0085	0.9964
5° 0'	1.8634	1.5707	1.4232	1.3337	1.2733	1.1959	1.1472	1.1124	1.0851	1.0616	1.0384	1.0248	1.0086	0.9962
10'	1.8920	1.5895	1.4371	1.3447	1.2823	1.2023	1.1519	1.1160	1.0878	1.0635	1.0396	1.0255	1.0088	0.9959
20'	1.9205	1.6083	1.4510	1.3557	1.2912	1.2086	1.1567	1.1196	1.0905	1.0654	1.0407	1.0262	1.0089	0.9957

* $\dfrac{\cos \theta}{\cos \theta_2}$ may be also written $\dfrac{\cos (\theta_2 + \beta - \gamma)}{\cos \theta_2}$

Interpolation for correction factors corresponding to values of ($\beta - \gamma$) lying between those given in the table may be made without error. Interpolation for correction factors corresponding to values of cos θ_2 lying between those given in the table may be made without exceeding an error of 0.0010 in the sections of the table lying between the heavy black lines; outside of these sections, and in all cases where the adjacent values of cos θ_2 are separated by the heavy black lines, the maximum error in interpolation will exceed 0.0010.

intersects the vertical line representing the power factor of the load. For example, the correction to be applied to the meter registration is shown to be –0.1% for 0.8 power factor lag (point H).

(b) Light Load. Using the average RCF value (1.004) and average phase-angle value (+6'), obtain the correction to be applied to the meter registration for light load following the same procedure as that for heavy load. For example, the correction to be applied to the meter registration is shown to be –0.3% for 0.8 power factor lag (point L).

NOTE: The nomograph in Fig C1 may be used when the ratio correction factors and phase angles are within ± 0.5% and ± 20 minutes of each other for: (1) The current transformers, or

(2) The voltage transformers.

For current transformers only, negative phase-angle values are found above the centerline, and positive phase-angle values below.

C2.2 Current and Voltage Transformers. For a transformer-rated meter with current and voltage transformers, the average combined transformer corrections may be obtained as follows:

(1) Calculate the average RCF of all current transformers connected to the meter for heavy load and for light load as in C2.1(1).

(2) Calculate the average RCF of the voltage transformers using values for the secondary voltage nearest to the voltage at the meter.

EXAMPLE 3:

	RCF
VT No 1	0.996
VT No 2	0.997
VT No 3	1.001
Average RCF	0.998

(3) Obtain a combined average RCF for heavy load and for light load by multiplying the average heavy load and the average light load RCFs of the current transformers by the average RCF of the voltage transformers.

EXAMPLE 4:

	Ratio Correction Factor	
	Heavy Load	Light Load
CT (from Example 1)	0.999	1.004
VT (from Example 3)	· 0.998	· 0.998
Combined average RCF	0.997	1.002

(4) Calculate the algebraic average phase angle of all current transformers connected to the meter for heavy load and light load as in C2.1(2).

(5) Calculate the algebraic average phase angle of the voltage transformers using values for the secondary voltage nearest to the voltage at the meter.

EXAMPLE 5:

	Phase Angle
VT No 1	– 3′
VT No 2	+15′
VT No 3	+12′
Average phase angle	+ 8′

(6) Obtain a combined average phase angle for heavy load and for light load by subtracting algebraically the average phase angle of the voltage transformers from the average heavy-load and average light-load phase angles of the current transformers.

EXAMPLE 6:

	Phase Angle	
	Heavy Load	Light Load
CT (from Example 2)	– 9′	+6′
VT (from Example 5)	(–) + 8′	(–) +8′
Combined average phase angle	–17′	–2′

(7) Obtain the heavy-load and the light-load average combined transformer corrections from Fig C1 as outlined in C2.1(3). For example, the line is now drawn between the combined average RCF of 0.997 and the combined average phase angle of –17 minutes for heavy load, and the correction to be applied to the meter registration is shown to be –0.1% for 0.8 power factor lag (point CH). For light load, the line is drawn between the combined average RCF of 1.002 and the combined average phase angle of –2 minutes, and the correction to be applied to the meter registration is shown to be –0.2% for 0.8 power factor lag (point CL).

NOTE: If the combined phase angle for current and voltage transformers (current transformer phase angle minus voltage transformer phase angle) results in a negative value, the figure is found above the center horizontal line; conversely, positive combined phase angles are found below the center horizontal line.

C2.3 Weighted Average Combined Transformer Correction. A weighted average combined transformer correction may be obtained in a manner comparable to that used for obtaining average percentage registration for watthour meters (see 6.1.8). The average combined transformer correction for heavy load and the average combined correction for light load would be added together and the sum divided by two in one method; and in another method, the heavy-load correction would be multiplied by four, added to the light-load correction, and the sum divided by five. The method used should be consistent with that adopted for meters.

An example of the application of final average transformer correction is as follows:

Meter registration (without correction)	= 99.2%
Standard watthour meter correction	= +0.1%
Final average combined transformer correction	= +0.6%
Final average meter registration	= 99.9%

NOTE: When the average combined transformer correction for light load differs from that at heavy load by more than 0.6%, it is advisable to apply the light-load and heavy-load average combined transformer corrections separately when calibrating the meter.

C3. Formulas

The formulas in this section may be employed to obtain the correction factors for use in testing meters independently of current and voltage transformers. The following symbols are used:

RCF_I = ratio correction factor of a current transformer

RCF_E = ratio correction factor of a voltage transformer

RCF_K = combined ratio correction factor = $RCF_I \cdot RCF_E$

β = phase angle of a current transformer; the angle between the primary current phasor and the secondary current phasor reversed

γ = phase angle of a voltage transformer; the angle between the primary voltage phasor and the secondary voltage phasor reversed

$\cos \theta$ = power factor of primary circuit

$\cos \theta_2$ = apparent power factor of load as measured on the secondary winding of the transformer or transformers

$PACF_I$ = phase-angle correction factor of a current transformer

$$= \frac{\cos (\theta_2 + \beta)}{\cos \theta_2}$$

$PACF_E$ = phase-angle correction factor of a voltage transformer

$$= \frac{\cos (\theta_2 - \gamma)}{\cos \theta_2}$$

$PACF_K$ = phase-angle correction factor for both a current and voltage transformer

$$= \frac{\cos (\theta_2 + \beta - \gamma)}{\cos \theta_2}$$

TCF = transformer correction factor for either current or voltage transformer = $RCF \cdot PACF$

FCF = final correction factor where both current and voltage transformers are used = $RCF_K \cdot PACF_K$

Referring to C2.2, an example is given of a transformer-rated meter with current and voltage transformers having light-load data of 1.002 for combined average RCF and –2 minutes for combined average phase angle $(\beta - \gamma)$.

For an apparent power factor of 0.8 lag, the phase-angle correction factor for the current and voltage transformers would be determined as follows:

$$PACF_K = \frac{\cos(\theta_2 + \beta - \gamma)}{\cos\theta_2} = \frac{\cos(36°52'12'' - 0°2'0'')}{0.8}$$

$$= \frac{\cos 36°50'12''}{0.8} = \frac{0.800348}{0.8} = 1.0004$$

The final correction factor (FCF) equals $RCF_K \cdot PACF_K = 1.002 \cdot 1.0004 = 1.0024$, so the correction to be applied to the meter would be -0.2%.

C4. Tables

Tables C1 and C2 show phase-angle correction factors at various power factors and are particularly useful when phase angles exceed the 20-minute values covered by the nomograph.

In C2.2, an example is given of a transformer-rated meter with current and voltage transformers having light-load data of 1.002 for combined average RCF and -2 minutes for combined average phase angle $(\beta - \gamma)$. For a power factor of 0.8 lag, we find from the first line of Table C2 that the phase-angle correction factor for -5 minutes and 0.8 power factor would be 1.0011. By interpolation, the phase-angle correction factor for -2 minutes would be 1.0004. The final correction factor (FCF) equals $RCF_K \cdot PACF_K = 1.002 \cdot 1.0004 = 1.0024$, so the correction to be applied to the meter registration would be -0.2%.

Appendix D
Pulse Devices

D1. General

The usual form of pulse initiator is that of an attachment to a watthour meter, so arranged that the number of pulses produced is proportional to the revolutions of the meter rotor. The operating principle may be mechanical, photoelectrical, magnetic, Hall effect, or other. The constants associated with pulse initiators are related mathematically, to allow the device to be checked with formulas in a manner similar to that used to check register gearing. The pulse-initiator output ratio is generally the only constant marked on the device by the manufacturer. This is sufficient to describe the end results, provided that the pulse initiator remains with the meter as originally purchased. Knowledge of intermediate gearing and related constants is required to check operation after repairs, part replacements, modifications, or conversions are made.

D2. Symbols

K_d, or KWC (demand constant of pulse receiver): The value of each received pulse divided by the demand interval, expressed in kilowatts per pulse, kilovars per pulse, or other suitable units

K_e, or KWHC (pulse-initiator output constant): The value of the measured quantity for each outgoing pulse of a pulse initiator, expressed in kilowatthours per pulse, kilovarhours per pulse, or other suitable units

K_h (watthour constant): The registration, expressed in watthours, corresponding to one revolution of the rotor of a watthour meter

M_p, or R/P (pulse-initiator output ratio): The number of meter rotor revolutions per output pulse of the pulse initiator

P_c (pulse-initiator coupling ratio): The number of revolutions of the pulse-initiating shaft for each output pulse

P_g (pulse-initiator gear ratio): The ratio of meter rotor revolutions to the revolutions of the pulse-initiating shaft

P_r (pulse-initiator ratio): The ratio of revolutions of the first gear of the pulse initiator to the revolutions of the pulse-initiating shaft

P_s (pulse-initiator shaft reduction): The ratio of meter rotor revolutions to the revolutions of the first gear of the pulse initiator

R_p: The ratio of input pulses to output pulses for totalizing relay

t (demand interval): The length of the interval of time upon which the demand measurement is based

D3. Pulse-Initiator Output Constant

The pulse initiator must be capable of providing a practical value of output constant (K_e or KWHC) for the total range of primary watthour constants possible. The optimum is usually to minimize K_e without exceeding the operational capacity of any single component of the installation. Modern demand equipment is characterized by larger pulse capacities and correspondingly smaller output constants for a given primary watthour constant. The pulse initiator, whether modern or otherwise, still provides the practical control for the dependent variable via the optional choice of output ratio M_p.

D4. Formulas

The demand constant (K_d or KWC) is of prime interest to the billing department of a utility. Its computation and verification are therefore of prime concern to the meterman. All factors that affect the K_d must be considered. Relays or totalizers that change the value of the pulse constant K_e by some ratio R_p must be included in the computations. The usual form of demand constant is in terms of active or reactive power per output pulse. This may lead to confusion, since it is energy (active or reactive) and not power alone that is required to produce pulses via motion of a meter rotor. The knowledge of interval t and its proper understanding, in formulas, is therefore important. In the following formulas, the interval t must be expressed in hours:

$$K_e = \frac{PK_h \cdot M_p}{1000}$$

$$K_d = \frac{K_e \cdot R_p}{t} \qquad (R_p = 1 \text{ when no totalizer is used})$$

$$M_p = \frac{1000 \cdot t \cdot K_d}{PK_h}$$

$$M_p = R/P = P_g \cdot P_c$$

$$P_g = P_s \cdot P_r$$

$$PK_h = K_h \cdot CTR \cdot VTR$$

where

CTR = current transformer ratio
VTR = voltage transformer ratio

Appendix E
Registering Mechanism and Meter Constants

E1. General

In 6.1, descriptions were given of methods for determining the accuracy of a meter as far as the speed of the rotor is concerned, and for physically checking register and gear ratios. It is equally important that it be determined mathematically that the relations between the register (dial) constant, watthour constant, register ratio, and gear ratio are correct. The register constant should always appear on the face of the register when other than one; the register ratio will be found marked on the register or on the nameplate; and the watthour constant usually will be found marked on the nameplate or on the disk. Manufacturers generally use one standard shaft reduction for all ratings of meters of the same type, but the information does not appear on the meter. The gear ratio is dependent on the shaft reduction and also the register ratio. The gear ratio information also does not appear on the meter.

E2. Symbols

K_h (watthour constant): The number of watthours per revolution of the meter rotor (disk)

K_r (register, or dial, constant): The multiplier used to convert the register reading to kilowatthours

R_g (gear ratio): The number of revolutions of the rotor (disk) for one revolution of the first dial pointer

R_r (register ratio): The number of revolutions of the first gear of the register for one revolution of the first dial pointer

R_s (shaft reduction): The number of revolutions of the meter rotor (disk) for one revolution of the first gear of the register

E3. Shaft Reduction

In some meters a single-pitch worm is used on the rotor, meshing with a worm wheel of 100 teeth on the register; thus, the shaft reduction is 100. A single-pitch worm is sometimes used with a 50-tooth worm wheel to give a shaft reduction of 50.

In others, a double-pitch worm is used on the rotor, meshing with a worm wheel of 100 teeth on the register; thus, the shaft reduction is 50.

In still others, pinions on the rotor meshing with gears on registers result in shaft reductions of $6^1/_4$, $8^1/_3$, etc.

Transfer gearing between the disk shaft and the register is used in a few types of meters. In some, it is of 1:1 ratio and has no effect on the shaft reduction. There are instances, however, where the transfer gearing is either $16^2/_3$ to 15 or $16^2/_3$ to 30.

The shaft reduction may be determined from the manufacturer's literature, from tables, by counting teeth in gears and pinions, or by test.

E4. Formulas

When the register constant (K_r), watthour constant (K_h), and shaft reduction (R_s) are known, the register ratio (R_r) may be determined by the following formula:

$$R_r = \frac{K_r \cdot 10 \cdot 1000}{K_h \cdot R_s}$$

For example, if $K_r = 1$, $K_h = 3.6$, and $R_s = 100$,

then

$$R_r = \frac{1 \cdot 10\,000}{3.6 \cdot 100} = 27^7/_9$$

Other useful formulas are as follows:

Registration of one revolution of first dial pointer

$$= \frac{K_h \cdot R_r \cdot R_s}{1000} = \text{kilowatthours}$$

$$R_g = R_r \cdot R_s$$

$$K_r = \frac{K_h \cdot R_r \cdot R_s}{10 \cdot 1000} = \frac{K_h \cdot R_g}{10\,000}$$

$$K_h = \frac{K_r \cdot 10 \cdot 1000}{R_r \cdot R_s}$$

$$R_s = \frac{K_r \cdot 10 \cdot 1000}{K_h \cdot R_r}$$

In the foregoing formulas, 10 is the numerical value of one revolution of the first dial pointer.

Appendix F
Standards for In-Service Performance — Direct-Current Watthour Meters

F1. General Requirements

Any meter that has an incorrect register constant, test constant, gear ratio, or dial train, is mechanically defective, or registers upon no load (creeps) shall not be placed in service or be allowed to remain in service without adjustment and correction.

F2. Accuracy Limits

Any meter that has an error in registration of more than ±2% at light load or at heavy load shall not be placed in service or allowed to remain in service without adjustment.

F3. Tests Prior to Installation

New meters or meters returned from service shall be inspected and tested prior to installation on customer's premises. Meters that require adjustment shall be adjusted as closely as practicable to the condition of zero error.

F4. Performance Tests

Before disturbing conditions that might affect its accuracy, the meter should be inspected and tested for creep and for accuracy, before removing the meter cover if practicable. Meters that require adjustment shall be adjusted as closely as practicable to the condition of zero error.

F5. Postinstallation Tests

The accuracy of direct-current meters is materially affected by adjacent magnetic fields; therefore, in addition to the regular tests made on such meters prior to installation, the meters should be tested within 60 days after installation on customer's premises.

F6. Periodic Tests

In order to determine their performance in service, direct-current meters should be tested in accordance with the following schedule:

Up to and including 8 kW: at least once in 3 years
Over 8 kW: at least once in every year

Appendix G
Historical Background

G1. Preface to the First Edition (1910)

In undertaking 2 years ago to formulate a Meter Code, it was the ambition of the Meter Committee of the Association of Edison Illuminating Companies (AEIC) to produce a reliable and up-to-date manual covering the many phases of electric meter practice as encountered by all companies, both large and small. It was the Committee's belief that such a Code, if intelligently prepared, would prove of great value not alone to those actually engaged in operating meters, but also to those interested in the practices of metering from other standpoints, namely, official, legal, etc. There was also felt an urgent need of a closer agreement between the manufacturers and the operating companies as to reasonable and satisfactory specifications covering both operation and design.

The development of such a Code with the collecting of the very large amount of necessary data was placed in the hands of the Electrical Testing Laboratories of New York, and at the Briarcliff Convention of 1909 there was presented the first issue of the Code, covering four sections and representing the first year's work. As a means of increasing the strength and support of the work, and at the same time avoiding duplication of effort along similar lines, it was arranged with the consent of the Executive Committee of both Associations to join hands with the Meter Committee of the National Electric Light Association (NELA) for the further development of the Code. The second year's work, therefore, represents the combined efforts of the Meter Committees of the two Associations.

The Code to date as here presented includes with minor revisions and corrections those sections which have been presented in the reports at the 1909 Edison Convention and the 1910 NELA Convention, and also two entirely new sections. It is hoped that it may find its place among the reliable books of reference in the hands of those responsible for, and interested in, the purchase, installation, and operation of electric meters.

A considerable amount of ground still remains to be covered, and it is only to be expected that, with changes and improvements in the art, revisions must from time to time become necessary, but it is the intention of the Committees to continue the work to its logical conclusion.

While the Code is naturally based upon scientific and technical principles, the commerical side of metering has been constantly kept in mind as of very great importance, and it is believed that due consideration has been given to this phase of the problem.

Although the work has been directed very closely by the two Committees, the burden of the undertaking has been carried by the Electrical Testing Laboratories, to which full credit should be given.

The Committees are indebted to Clayton H. Sharp for his personal interest and cooperation in the conduct of the work and to W. W.

Crawford, also of the Laboratories, for the zeal and discrimination which he has displayed in preparing the drafts of the Code for the Committees' consideration.

The Committees would also acknowledge most gratefully the hearty and valuable cooperation of the manufacturing companies and particularly that of F. P. Cox and L. T. Robinson of the General Electric Company, and William Bradshaw of the Westinghouse Electric and Manufacturing Company. It is the earnest desire of the Committees that the Code may prove its value to all of those interested in the precise commercial measurement of electrical energy and may contribute to the advancement of the art.

Committee personnel:

AEIC	NELA
J. W. Cowles, Chairman	G. A. Sawin, Chairman
O. J. Bushnell	W. H. Fellows
George Ross Green	W. E. McCoy
J. T. Hutchings	
S. G. Rhodes	

G2. Preface to the Second Edition (1922)

This edition of the Code for Electricity Meters is a revised and complete compilation of the sections issued separately during the past 5 years. The revision and arrangement here have been under the supervision of the Meter Committees of the Association of Edison Illuminating Companies and the National Electric Light Association.

Advantage was taken of the printing of the Code in complete form to make such revisions in the text and to add such new matter as appeared desirable. The Electrical Testing Laboratories joined with the Committees in this revision and compilation as they did in the original preparation of the various sections of the Code, and this revised edition has their approval.

The Code for Electricity Meters has been generally accepted as a standard of reference for meter practice. Its revision, completion, and appearance in one volume enhance its value for this purpose.

Committee personnel:

AEIC	NELA
S. G. Rhodes, Chairman	O. J. Bushnell, Chairman
O. J. Bushnell	W. H. Fellows
J. W. Cowles	W. E. McCoy
George Ross Green	F. A. Vaughn
J. T. Hutchings	W. L. Wadsworth
G. A. Sawin	

G3. Preface to the Third Edition (1928)

This edition of the Code for Electricity Meters is a completely revised and rearranged compilation of the Second Edition, issued in 1912, and the section on demand meters, issued in 1920. The 1912–20 edition of the Code was approved as an American Standard by the

American Engineering Standards Committee in July 1922 (C12-1922). The present revision has been made under the joint sponsorship of the Association of Edison Illuminating Companies, the National Electric Light Association, and the US National Bureau of Standards by a Sectional Committee representing all interested organizations, in accordance with the procedure established by the American Engineering Standards Committee for the revision of American Standards.

The sponsors hereby express their appreciation to the members of the Sectional Committee and their associates for the painstaking and careful manner in which the revision was carried out.

A preliminary draft was presented at a meeting of the Sectional Committee on April 1, 1926. This draft was approved in general outline, and referred to an editorial committee consisting of Messrs Brooks, Currier, Doyle, Fellows, Hill, Koenig, Meyer, and Pratt. This committee carefully reviewed the draft, agreed upon a standard form and arrangement, and appointed H. Koenig, the Secretary of the Sectional Committee, to prepare the final draft for the printer. A considerable amount of material appearing in the Second Edition has been omitted, particularly the circuit diagrams in Section VII, all of Section IX, and the maintenance paragraphs of Section X. The sections or chapters have been renumbered. The omitted material is fully covered in the *Handbook for Electrical Metermen*, where it now properly belongs.

This Code, as revised, was submitted in galley-proof form to all the members of the Sectional Committee for final approval by letter ballot, and it was then formally approved by each of the sponsors. The sponsors, acting jointly, presented the Code to the American Engineering Standards Committee for approval as American Standard, and it was so approved February 20, 1928.

Committee personnel:

W. M. Bradshaw	F. Holmes	J. Franklin Meyer
O. J. Bushnell	F. A. Kartak	A. L. Pierce
F. P. Cox	H. Koenig	G. A. Sawin
Burleigh Currier	R. C. Lanphier	C. H. Sharp
E. D. Doyle	F. V. Magalhaes	C. R. Vanneman
R. W. Eaton	Alexander Maxwell	F. G. Vaughen
C. W. Hayden		W. L. Wadsworth

The Sectional Committee was formally organized March 14, 1924; J. Franklin Meyer, Chairman; E. D. Doyle, Secretary, later succeeded by H. Koenig.

The actual revision of the Code was done by four technical subcommittees, as authorized by the Sectional Committee. These subcommittees were:

(1) Acceptance Specifications: F. V. Magalhaes, Chairman; A. J. Allen, W. M. Bradshaw, H. B. Brooks, O. J. Bushnell, C. J. Clarke, C. I. Hall, F. C. Holtz, C. H. Ingalls, A. E. Knowlton, W. H. Pratt

(2) Installation and Maintenance Methods: B. Currier, Chairman; A. S. Albright, A. J. Allen, W. H. Fellows, R. C. Fryer, E. E. Hill,

C. H. Ingalls, A. G. Turnbull, W. L. Wadsworth

(3) Standards: E. D. Doyle, Chairman; A. S. Albright, C. J. Clarke, H. G. Hamann, E. E. Hill

(4) Definitions: J. F. Meyer, Chairman; W. H. Fellows, F. C. Holtz, F. A. Kartak, C. H. Sharp

G4. Preface to the Fourth Edition (1941)

This fourth edition of the Code for Electricity Meters was prepared by Sectional Committee C12 of the American Standards Association. The sponsors were the National Bureau of Standards, and the American Standards Association — Electric Light and Power Group (the Association of Edison Illuminating Companies and the Edison Electric Institute).

The Sectional Committee C12 which prepared the revision was as follows:

J. Franklin Meyer, *Chairman*
H. C. Koenig, *Secretary*

A. J. Allen	E. E. Hill
R. D. Bennett	P. L. Holland
W. M. Bradshaw	F. C. Holtz
H. B. Brooks	R. E. Johnson
J. O'R. Coleman	N. S. Meyers
O. K. Coleman	R. H. Nexsen
Stanley S. Green	W. H. Pratt
C. B. Hayden	W. C. Wagner

The work of revision was divided into six major sections and was done by the following six subcommittees:

(1) Definitions: R. D. Bennett, Chairman; H. B. Brooks, P. G. Elliott, W. H. Fellows, R. E. Johnson, E. E. Kline, W. H. Pratt

(2) Standards and Metering: H. B. Brooks, Chairman; A. S. Albright, W. M. Bradshaw, F. E. Davis, Jr, F. C. Holtz, H. C. Koenig, G. R. Sturtevant

(3) Specifications for Acceptance of Types of Electricity Meters and Auxiliary Devices: W. C. Wagner, Chairman; W. M. Bradshaw, H. B. Brooks, A. B. Craig, W. R. Frampton, E. E. Hill, H. C. Koenig, R. H. Nexsen, W. H. Pratt

(4) Installation Methods and Watthour Meter Test Methods: O. K. Coleman, Chairman; A. P. Good, Stanley S. Green, C. B. Hayden, N. S. Meyers, L. D. Price

(5) Laboratory and Service Tests: P. L. Holland, Chairman; J. S. Cruikshank, P. G. Elliott, J. H. Goss, E. E. Hill, J. C. Langdell, F. L. Pavey

(6) Demand Meters: A. J. Allen, Chairman; F. C. Holtz, R. E. Johnson, E. A. LeFever, R. H. Nexsen, A. R. Rutter, W. C. Wagner, W. H. Witherow

G5. Preface to the Fifth Edition (1965)

Following the issuance of the Fourth Edition of the Code for Electricity Meters in 1941, a modification of periodic test schedules (Paragraph 827 — changing the test period to 96 months for ac meters rated to 12 kVA) was issued as an American War Standard on November 5, 1942. This change was approved by ASA as an American Standard in 1947. In 1957, American Standard Code for Electricity Meters, C12-1947, and American Standard Revisions to the Code for Electricity Meters, C12a-1947, were reaffirmed.

Many improvements and innovations in meters and their auxiliary equipment, and in metering practices, have taken place since the Fourth Edition of this Code was issued. These developments were taken into account in preparing the present edition. For the first time, the Code recognizes that statistical methods may be applied to in-service testing of meters to reveal where testing and maintenance effort should be directed; and guidance is offered toward the selection of sound statistical procedures. The other sections of the Code have also been broadened and largely rewritten to cover other phases of electricity metering in line with the present state of the art.

Finally, it should be noted that the name of this standard has been changed to American Standard Code for Electricity Metering, as the committee believed that this title more accurately described the content of the standard.

This edition of the American Standard Code for Electricity Metering was prepared by Sectional Committee C12 of the American Standards Association. The sponsors are the National Bureau of Standards and the Edison Electric Institute.

The personnel of Sectional Committee C12 that prepared this revision of the Code were as follows:

F. K. Harris, *Chairman*
A. T. Higgins, *Secretary*

J. Anderson	R. E. Purucker
T. D. Barnes	A. W. Rauth
D. T. Canfield	R. A. Road
W. C. Downing, Jr	G. B. M. Robertson
J. W. Dye	F. H. Rogers
P. W. Hale	R. S. Smith
H. H. Hunter	L. O. Steger
H. W. Kelley	G. P. Steinmetz
J. D. McKechnie	T. P. Wadsworth
W. J. Piper	G. J. Yanda

Liaison Members: L. V. Hunt
J. M. Vanderleck

The work of revision was done by a number of task forces, and was reviewed by the Sectional Committee. These task forces and their assignments were as follows:

(1) Definitions: W. J. Piper
(2) Measurement of Power and Energy: D. T. Canfield
(3) Standards: F. K. Harris, Chairman; E. F. Blair
(4) Acceptance of New Types of Meters: G. B. M. Robertson, Chairman; T. D. Barnes, E. F. Blair, J. D. McKechnie, R. A. Road, R. S. Smith
(5) Watthour Meter Test Methods: P. W. Hale, Chairman; J. Ander-

son, T. D. Barnes, W. C. Downing, Jr, H. W. Kelley, J. D. McKechnie, E. C. Nuesse, R. A. Road, F. H. Rogers

(6) Installation Requirements: H. W. Kelley, Chairman; E. B. Hicks, H. H. Hunter, L. H. Keever, R. E. Purucker, A. W. Rauth, L. O. Steger

(7) Instrument Transformers and Auxiliary Devices: J. W. Dye, Chairman; E. F. Blair, F. R. D'Entremont, B. L. Dunfee, W. H. Farrington, H. W. Kelley

(8) In-Service Tests of Watthour Meters: H. H. Hunter, Chairman; F. K. Harris, A. L. Hobson, C. L. Lucal, J. D. McKechnie, C. V. Morey, R. E. Purucker, F. H. Rogers, L. O. Steger, G. Wey

(9) Demand Meters (Acceptance, Test Methods, In-Service Tests): G. J. Yanda, Chairman; R. V. Adams, W. C. Downing, Jr, P. W. Hale, F. M. Hoppe, W. J. Piper, R. A. Road, R. J. Stowel

(10) Editorial: G. B. M. Robertson, Chairman; J. Anderson, P. W. Hale, F. K. Harris, A. T. Higgins, H. H. Hunter, H. W. Kelley, F. H. Rogers, G. J. Yanda

G6. Preface to the Sixth Edition

A number of significant advances have been made in the design of watthour meters, in the verification of their accuracy, and in demand metering, since the Fifth Edition of this Code was issued in 1965. Improvements in bearings and mechanical construction, and new sealing techniques that exclude dust have made modern meters remarkably stable, as well as accurate. Sampling methods of in-service testing sanctioned by the 1965 Code have been shown to be economical and effective.

A new form of auxiliary device, known as a pulse recorder, has come into general use during the past 10 years. It records, on magnetic or paper tape, pulses received from pulse initiators installed on watthour or other integrating meters. The tapes are processed by automated equipment using computer techniques, thus reducing human errors and speeding up accounting and data-interpretation processes for both customer billing and survey installations.

These developments as well as others have been taken into account in this edition of the Code. Recommended periodic test intervals for modern meters have been lengthened, and sampling methods have been extended to additional kinds of meters. In addition, performance requirements have been incorporated for the new types of pulse devices and for the standard watthour meters used as references to maintain the kilowatthour or to test other meters. Many other changes have been made.

This standard is a revision of American National Standard Code for Electricity Metering, C12-1965. The secretariat of American National Standards Committee C12 is held by the Edison Electric Institute and the National Bureau of Standards.

This standard was processed and approved for submittal to ANSI by the American National Standards Committee on Code for Electricity Metering, C12. Committee approval of the standard does not necessarily imply that all committee members voted for its approval. At the time it approved this standard, the C12 Committee had the following members:

F. L. Hermach, *Chairman*
A. T. Higgins, *Secretary*

J. Anderson	A. Fini	T. J. Pearson
J. C. Arnold*	L. M. Holdaway†	C. F. Riederer
E. L. Barker	B. E. Kibbe	C. Ringold
D. B. Berry	F. G. Kuhn	R. A. Road
E. F. Blair	F. J. Levitsky	E. W. Schwarz
H. L. Colbeth	D. McAuliff	L. O. Steger§
C. R. Collinsworth	W. E. Osborn	J. M. Vanderleck

The work of revision was done by a number of subcommittees, and was reviewed by the C12 Standards Committee. The assignments of these subcommittees were as follows:

(1) Definitions, and (2) Measurement of Power and Energy: E. F. Blair, Chairman; R. S. Turgel, J. M. Vanderleck, A. Yenkelun

(3) Standards and Standardizing Equipment: F. L. Hermach, Chairman; M. F. Borleis, W. E. Osborn, J. Roth, E. W. Schwarz, D. M. Smith

(4) Acceptance of New Types of Watthour Meters: A. Fini, Chairman; J. Anderson, D. B. Berry, E. F. Blair, M. F. Borleis, C. R. Collinsworth, F. G. Kuhn, D. McAuliff, G. F. Walsh

(5) Watthour Meter Test Methods: F. J. Levitsky, Chairman; J. Anderson, E. F. Blair, T. J. Pearson

(6) Installation Requirements: B. E. Kibbe, Chairman; D. Berry, A. Browne, M. A. Frederickson, L. M. Holdaway, H. W. Redecker

(7) Instrument Transformers and Auxiliary Devices: T. J. Pearson, Chairman; B. L. Dunfee, F. A. Fragola, J. Landry, R. Stetson

(8) In-Service Tests of Watthour Meters: H. L. Colbeth, Chairman; E. L. Barker, M. A. Frederickson, J. Keever, J. C. Liewehr, B. Renz, C. F. Riederer, J. Suridis

(9) Demand Meter and Pulse Devices: C. R. Collinsworth, Chairman; E. C. Benbow, H. A. Duckworth, R. Hopkins, S. C. McCollum, C. F. Riederer, C. Ringold, R. J. Stowell, G. F. Walsh

(10) Editorial: R. A. Road, Chairman; J. Anderson, F. L. Hermach, A. T. Higgins, F. J. Levitsky, W. E. Osborn, C. F. Riederer

---

*Replaced Mr Holdaway in September, 1974.
†Deceased, July, 1974.
§ Until retirement in April, 1973.

Index

206